NANOTECHNOLOGY SCIENCE AND TECHNOLOGY

MODERN NANOCHEMISTRY

NANOTECHNOLOGY SCIENCE AND TECHNOLOGY

Additional books in this series can be found on Nova's website under the Series tab.

Additional E-books in this series can be found on Nova's website under the E-books tab.

CHEMISTRY RESEARCH AND APPLICATIONS

Additional books in this series can be found on Nova's website under the Series tab.

Additional E-books in this series can be found on Nova's website under the E-books tab.

NANOTECHNOLOGY SCIENCE AND TECHNOLOGY

MODERN NANOCHEMISTRY

A. K. HAGHI

AND

G. E. ZAIKOV

EDITORS

Nova Science Publishers, Inc.

New York

NOTICE TO THE READER

The Publisher has taken reasonable care in the preparation of this book, but makes no expressed or implied warranty of any kind and assumes no responsibility for any errors or omissions. No liability is assumed for incidental or consequential damages in connection with or arising out of information contained in this book. The Publisher shall not be liable for any special, consequential, or exemplary damages resulting, in whole or in part, from the readers' use of, or reliance upon, this material. Any parts of this book based on government reports are so indicated and copyright is claimed for those parts to the extent applicable to compilations of such works.

Independent verification should be sought for any data, advice or recommendations contained in this book. In addition, no responsibility is assumed by the publisher for any injury and/or damage to persons or property arising from any methods, products, instructions, ideas or otherwise contained in this publication.

This publication is designed to provide accurate and authoritative information with regard to the subject matter covered herein. It is sold with the clear understanding that the Publisher is not engaged in rendering legal or any other professional services. If legal or any other expert assistance is required, the services of a competent person should be sought. FROM A DECLARATION OF PARTICIPANTS JOINTLY ADOPTED BY A COMMITTEE OF THE AMERICAN BAR ASSOCIATION AND A COMMITTEE OF PUBLISHERS.

Additional color graphics may be available in the e-book version of this book.

Library of Congress Cataloging-in-Publication Data

Modern nanochemistry / [edited by] A.K. Haghi, G.E. Zaikov.
    p. cm.
  Includes index.
  ISBN 978-1-61209-992-7 (softcover)
  1. Drug delivery systems. 2. Nanoparticles. I. Haghi, A. K. II. Zaikov,
G. E. (Gennadii Efremovich), 1935-
  RS201.N35M63 2011
  615'.6--dc22
                              2011007123

Published by Nova Science Publishers, Inc. ✛ *New York*

This volume is dedicated to the memory of Frank Columbus

On December 1st 2010, Frank H. Columbus Jr. (President and Editor-in-Chief of Nova Science Publishers, New York) passed away suddenly at his home in New York.

We lost our colleague, our good friend, a nearly perfect person who helped scientists from all over the world. Particularly Frank did much for the popularization of Russian and Georgian scientific research, publishing a few thousand books based on the research of Soviet (Russian, Georgian, Ukranian etc.) scientists.

Frank was born on February 26th 1941 in Pennsylvania. He joined the army upon graduation of high school and went on to complete his education at the University of Maryland and at George Washington University. In 1969, he became the Vice-President of Cambridge Scientific. In 1975, he was invited to work for Plenum Publishing where he was the Vice-President until 1985, when he founded Nova Science Publishers, Inc.

Frank Columbus did a lot for the prosperity of many Soviet (Russian, Georgian, Ukranian, Armenian, Kazakh, Kyrgiz, etc.) scientists publishing books with achievements of their research. He did the same for scientists from East Europe – Poland, Hungary, Czeckoslovakia (today it is Czeck republic and Slovakia), Romania and Bulgaria.

He was a unique person who enjoyed studying throughout the course of his life, who felt at home in his country which he loved and was proud of, as well as in Russia and Georgia.

There is a famous Russian proverb: "The man is alive if people remember him." In this case, Frank is alive and will always be in our memories while we are living. He will be remembered for his talent, professionalism, brilliant ideas and above all – for his heart.

Professor Gennady Efremovich Zaikov
Honoured Member of Russian Science
Head of Polymer Division, IBCP

CONTENTS

PREFACE

Nanochemistry is a new discipline concerned with the unique properties associated with assemblies of atoms or molecules on a scale between that of the individual building blocks and the bulk material. At this level, quantum effects can be significant, and also innovative ways of carrying out chemical reactions become possible. It is the science of tools, technologies, and methodologies for chemical synthesis, analysis, and biochemical diagnostics, performed in nanoliter to femtoliter to domains.

Nanochemistry is the use of synthetic chemistry to make nanoscale building blocks of desired shape, size, composition and surface structure, charge and functionality with an optional target to control self-assembly of these building blocks at various scale-lengths.

Nanochemistry use semi-conductors that only conduct electricity in specific conditions. As the semi-conductors are much smaller than normal conductors the product can be much smaller.

Modern concepts in nanochemistry is strongly interrelated cutting edge frontiers in research in the chemical sciences. The results of recent work in the area are now an increasing part of modern degree courses and hugely important to researchers.

Modern concepts in nanochemistry clearly outline the fundamentals that underlie supramolecular chemistry and nanochemistry and take an umbrella view of the whole area. This concise textbook traces the fascinating modern practice of the chemistry of the non-covalent bond from its fundamental origins through to it expression in the emergence of nanochemistry.

Fusing synthetic materials and supramolecular chemistry with crystal engineering and the emerging principles of nanotechnology, the book is an ideal introduction to current chemical thought for researchers and a superb resource for students entering these exciting areas for the first time. The book builds from first principles rather than adopting a review style and includes key references to guide the reader through influential work.

A. K. Haghi, University of Guilan, Iran
G.E. Zaikov, Russian Academy of Sciences

In: Modern Nanochemistry   ISBN: 978-1-61209-992-7
Eds: A. K. Haghi and G. E. Zaikov   © 2011 Nova Science Publishers, Inc.

Chapter 1

DESIGN AND SYNTHESIS OF DRUG DELIVERY SYSTEMS BASED ON CHITOSAN NANOPARTICLES

M. R. Saboktakin[1] and A. K. Haghi[2]†*
[1]Department of Nanotechnology, Baku State University, Azerbaijan
[2]University of Guilan, Iran

ABSTRACT

The main objective of this research is to design a new extended release multiparticulate delivery system by incorporation into nanoparticles made of chitosan-polymethacrylic acid copolymers. As the first part of a continued research on conversion of chitosan to useful biopolymer-based materials, by grafting polymethacrylic acid (PMAA), free radical graft copolymerization was carried out at 70 °C, with bisacrylamide as a cross-linking agent and persulfate as an initiator. Equilibrium swelling studies were carried out in enzyme-free simulated gastric and intestinal fluids. Also, the paclitaxel as a model drug was entrapped in these nano-gels and in vitro release profiles were established separately in both enzyme-free SGF and SIF. The drug release was found to be faster in SIF.

---

* saboktakin123@yahoo.com
† Haghi@Guilan.ac.ir

INTRODUCTION

Natural polymers have potential pharmaceutical applications because of their low toxicity, biocompatibility, and excellent biodegradability. In recent years, biodegradable polymeric systems have gained importance for design of surgical devices, artificial organs, drug delivery systems with different routes of administration, carriers of immobilized enzymes and cells, biosensors, ocular inserts, and materials for orthopedic applications (BrOndsted & Kope˘cek, 1990). These polymers are classified as either synthetic (polyesters, polyamides, polyanhydrides) or natural (polyamino acids, polysaccharides) (Giammona, Pitarresi, Cavallora, & Spadaro, 1999; Krogars et al., 2000). Polysaccharide-based polymers represent a major class of biomaterials, which includes agarose, alginate, carageenan, dextran, and chitosan. Chitosan [_(1, 4)2- amino-2-d-glucose] is a cationic biopolymer produced by alkaline N-deacetylation of chitin, which is the main component of the shells of crab, shrimp, and krill (Chiu, Hsiue, Lee, & Huang, 1999; Jabbari & Nozari, 2000). Chitosan is a functional linear polymer derived from chitin, the most abundant natural polysaccharide on the earth after cellulose, and it is not digested in the upper GI tract by human digestive enzymes (Fanta & Doane, 1986; Furda, 1983). Chitosan is a copolymer consisting of 2-amino-2-deoxyd- glucose and 2-acetamido-2-deoxy-d-glucose units links with β-(1-4) bonds. It should be susceptible to glycosidic hydrolysis by microbial enzymes in the colon because it possesses glycosidic linkages similar to those of other enzymatically depolymerized polysaccharides. Among diverse approaches that are possible for modifying polysaccharides, grafting of synthetic polymer is a convenient method for adding new properties to a polysaccharide with minimum loss of its initial properties (Saboktakin, Maharramov, & Ramazanov, 2007; Peppas, 1987). Graft copolymerization of vinyl monomers onto polysaccharides using free radical initiation, has attracted the interest of many scientists. Up to now, considerable works have been devoted to the grafting of vinyl monomers onto the substrates, especially starch and cellulose (Jabbari & Nozari, 2000; Xu & Li, 2005). Existence of polar functionally groups as carboxylic acid need not only for bioadhesive properties but also for

pH-sensitive properties of polymer (Ratner, 1989; Thierry, Winnik, Mehri, & Tabrizian, 2003), because the increase of MAA content in the hydrogels provides more hydrogen bonds at low pH and more electrostatic repulsion at high pH. It is as a part of our research program on chitosan modification to prepare materials with pHsensitive properties for use as drug delivery

(Mahfouz, Hamm, & Taupitz, 1997; Schmitz et al., 2000; Bloembergen & Pershan, 1967).

In this study, we hypothesized that the absorption of paclitaxel could be enhanced by administration with chitosan-polymethacrylic acid nanoparticles because of their greater per- meability properties (Puttpipatkhachorn, Nunthanid, & Yamamato, 2001).

MATERIALS AND METHODS

Chitosan with 1:1 molar ratios of methacrylic acid were polymerized at 60–70 °C in a thermostatic water bath, bis-acrylamide as a cross-linking agent (CA), using persulfate as an initiator ([I] = 0.02M) and water as the solvent (50 mL). The polymeric system was stirred by mechanical stirrer to sticky nanoparticles and it was separated from medium without solvent addition. All the experiments were carried out in Pyrex glass ampoules. After the specific time (48 h), the precipitated network polymer was collected and dried in vacuum. Chitosan-methacrylic acid copolymer suspensions of 0.2% (w/v) were prepared in 1% acetic acid. Sodium tripolyphosphate (TPP, 1.0%) was added dropwise to 6mL of chitosan with stirring, followed by sonication with a dismembrator for 10 s at a power setting of 3W.The resulting chitosan particle suspension was centrifuged at 10, 000×g for 10 min.The pelleted particles were resuspended in deionized water with 10 s sonication and lyophilized. The mean size and zeta potential of the chitosan-methacrylic acid nanoparticles were determined by photon correlation spectroscopy using ZetaPlus particle analyzer. The major amount of drug was adsorbed by 2mg of chitosan nanoparticles in a certain time period. Chitosan-methacrylic acid nanoparticle suspensions (4 mg/mL) were mixed with paclitaxel solutions (0.5 and 1mg/mL), vortexed, and incubated at 37 °C for 1, 6, 12 and 18 h. After adsorption, the suspensions were centrifuged at 10, 000×g for 10 min and free drug was measured in the supernatant by a colorimetric method using periodic acid/Schiff (PAS) staining. Schiff reagent was prepared by diluting pararosaniline solution (40 g/L in 2MHCl, Sigma) with water to give a final concentration of 1.0% sodium bisulfite (80 mg) was added to 5mL of Schiff reagent and the resultant solution was incubated at 37 °C until it became colorless or pale yellow. Periodic acid solution was freshly prepared by adding 10_L of 50% periodic acid to 7mL of 7% acetic acid. Supernatants were mixed with 100_L of dilute periodic acid and incubated for 2 h at 37 °C. Then, 100_L of Schiff reagent was added at room temperature, and after 30 min the absorbance was measured at 560 nm.

The following procedure was used to assess the stability of paclitaxel during the bead preparation process. The prepared nanoparticles were extracted twice with a solvent mixture of 1:1 acetonitrile and ethanol (v/v), the extract was evaporated, the residue was injected onto HPLC column. Stability-indicating chromatographic method was adopted for this purpose. The method consisted of aSymmetryC18 column (254mm×4.6mm; 5_m)run using a mobile phase of composition methanol:water (70:30, v/v) at a flow rate of 0.5 mL/min, a waters pump (600E), and eluants. A definite weight range of 10–15mg of nanoparticles were cut and placed in a 1.5mL capacity microcentrifuge tube containing 1mL of release medium of the following composition at 37 °C: phosphate buffered saline (140mM, pH7.4) with 0.1% sodium azide and 0.1% Tween 80. At predetermined time points, 100_L of release medium was sampled with replacement to which 3mL of scintillation cocktail was added and vortexed before liquid scintillation counting. The cumulative amount of paclitaxel released as a function of time was calculated. To study the molecular properties of paclitaxel and chitosan- PMAA, the solid-state characterization was done by the application of thermal, X-ray diffraction, and microscopy techniques. During these studies, the solid characteristics of paclitaxel and chitosan- PMAA were compared with those of nanoparticles to reveal any changes occurring as a result of nanoparticle preparation. Differential scanning calorimetry (DSC) studies were performed with a Mettler Toledo 821 thermal analyzer (Greifensee, Switzerland) calibrated with indium as standard. For thermogram acquisition, sample sizes of 1–5mg were scanned with a heating rate of 5 °C/min over a temperature range of 25–300 °C. In order to check the reversibility of transition, samples were heated to a point just above the corresponding transition temperature, cooled to room temperature, and reheated up to 300 °C. Paclitaxel samples and chitosan-PMAA beads were viewed using a Philips XL-30 E SEM scanning electron microscope (SEM) at 30 kV (max.) for morphological examination. Powder samples of paclitaxel and beads were mounted onto aluminium stubs using double-sided adhesive tape and then sputter coated with a thin layer of gold at 10 Torr vacuum before examination. The specimens were scanned with an electron beam of 1.2 kV acceleration potential, and images were collected in collected in secondary electron mode. Molecular arrangement of paclitaxel and chitosan-PMAA in powder as well as in nanoparticles were compared by powder X-ray diffraction patterns acquired at room temperature on a Philips PW 1729 diffractometer (Eindhoven, Netherlands) using CuK radiation.

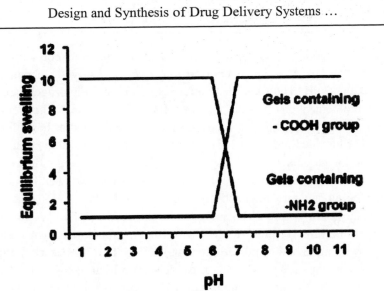

Figure 1. Equilibrium degree of swelling in response to pH.

X-ray diffraction is a proven tool to study crystal lattice arrangements and yields very useful information on degree of sample crystallinity.

RESULTS

In the present study, nanoparticles were prepared by the classical method, which involves spreading a uniform layer of polymer dispersion followed by a drying step for removal of solvent system. Since bead preparation methodology involved a heating step, it may have had a detrimental effect on the chemical stability of drug. Hence, stability assessment of paclitaxel impregnated in bead was done using stability-indicating method. For this purpose, paclitaxel was extracted from bead and analyzed by HPLC. A single peak at 21.2 min representing paclitaxel (with no additional peaks) was detected in the chromatogram, suggesting that the molecule was stable during preparation of beads Paclitaxel was extracted from different regions of chitosan- PMAA nanoparticles using acetonitrile:ETOH (1:1, v/v) solvent system. After normalization of amount of paclitaxel on weight basis of nanoparticles, the results indicated that the variation in distribution of paclitaxel in different regions of nanoparticles were <16%. The composition of the polymer defines its nature as a neutral or ionic network and furthermore, its hydrophilic/hydrophobic characteristics. Ionic hydrogels, which could be cationic, containing basic functional groups or anionic, containing acidic functional groups, have been reported to be very sensitive to changes in the environmental pH. The swelling properties of the ionic hydrogels are unique due to the ionization of their pendent functional groups. The equilibrium swelling behavior of ionic hydrogels containing acidic and/or basic functional groups is illustrated in Figure 1. Hydrogels containing basic functional groups are found increased swelling activity in acidic conditions and reduced in basic conditions but pH-sensitive anionic hydrogels shows low swelling activity in acidic medium and very high activity in basic medium.

As shown in Figure 2, an increase in the content of MAA in the freed monomer mixtures resulted in less swelling in simulated gastric fluid but greater swelling in and simulated intestinal fluids. This is because the increase of MAA content in the hydrogels provides more hydrogen bonds at low pH and more electrostatic repulsion at high pH. Figure 4 shows scanning electron microscope (SEM) of graft chitosan copolymer with polymethacrylic acid and nano-polymer bonded drug. Nano- and micro-polymer bonded drugs (50 mg) were poured into 3mL of aqueous buffer solution (SGF: pH 1 or SIF: pH 7.4). The mixture was introduced into a cellophane membrane dialysis bag. The bag was closed and transferred to a flask containing 20mL of the same solution maintained at 37 °C. The external solution was continuously stirred, and 3mL samples were removed at selected intervals. The removed volume was replaced with SGF or SIF. Triplicate samples were used. The sample of hydrolyzate was analyzed by UV spectrophotometer, and the quantity of paclitaxel was determined using a standard calibration curve obtained under the same conditions (Figure 3). It appears that the degree of swelling depends on their particle size. As shown in Figure 2, a decrease in the molecular size of carriers increased the swelling rate. The thermal behavior of a polymer is important in relation to its properties for controlling the release rate in order to have a suitable drug dosage form. The glass transition temperature (Tg) was determined from the DSC thermograms. The higher Tg values probably related to the introduction of cross-links, which would decrease the flexibility of the chains and the ability of the chains to undergo segmental motion, which would increase the Tg values. On the other hand, the introduction of a strongly polar carboxylic acid group can increase the Tg value because of the formation of internal hydrogen bonds between the polymer chains (Figure 4). X-ray diffraction is also used to study the degree of crystallinity of pharmaceutical drugs and excipients. A lower 2Θ value indicates larger d-spacings, while an increase in the number of high-angle reflections indicates higher molecular state order.

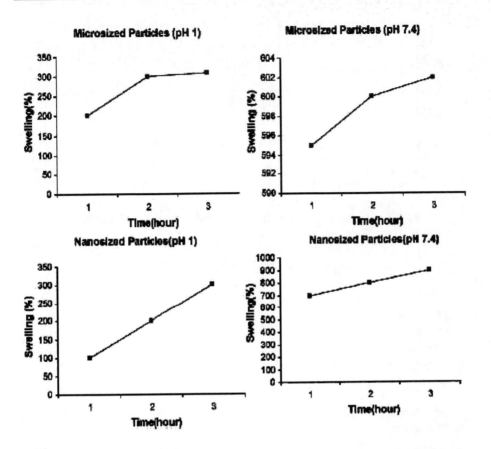

Figure 2. Time-dependent swelling behavior of micro- and nano-carriers for paclitaxel drug model as a fuction of time at 37°C.

In addition, broadness of reflections, high noise, and low peak intensities are characteristics of a poorly crystalline material. A broad hump in the diffraction pattern of chitosan hydogel extending over a large range of 2Θ suggests that chitosan is present in amorphous state in the film.

X-ray diffraction patterns of paclitaxel and chitosan- PMAA hydrogel film were obtained and compared, which revealed marked differences in the molecular state of paclitaxel (Figure 5). X-ray diffractogram of paclitaxel and chitosan-PMAA hydrogel film shows several high-angle diffraction peaks were observed at the following 2Θ values: 24.1, 27.4, 29.2, 36.4, 40.3, and 44.6°. The 29.6° 2Θ peak had the highest intensity as observed for hydrogel film.

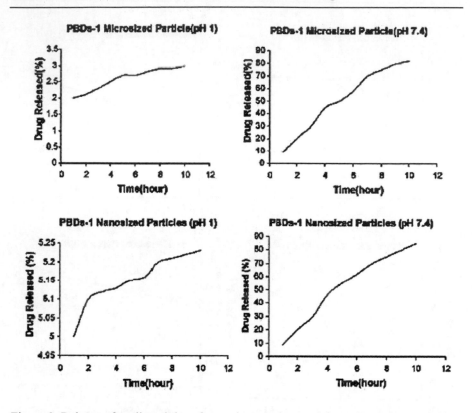

Figure 3. Release of paclitaxel drug from micro- and nano-polymeric carriers as a function of time at 37°C.

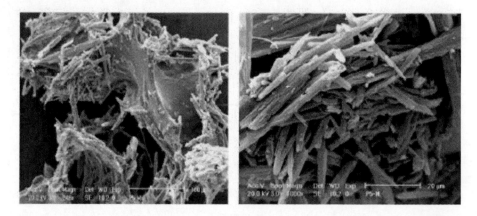

Figure 4. SEM of paclitaxel- chitosan nanoparticles.

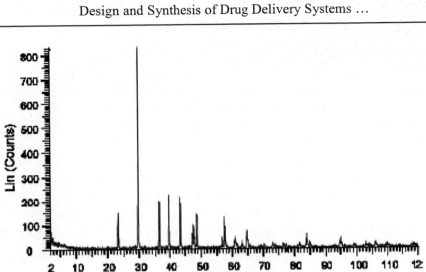

Figure 5. XRD pattern of chitosan nanoparticles.

DISCUSSION

In order to develop a local delivery system for paclitaxel, a biodegradable N-chitosan-PMAA bead was attempted, which to our knowledge in the first effort of its kind. Potential formulation problems were anticipated since chitosan is only soluble in aqueous acidic solutions, whereas paclitaxel, being a hydrophobic drug, is insoluble under similar conditions. In early stages of ormula optimization studies, it was observed that paclitaxel was incorporated into bead. The primary mechanisms for release of drugs from matrix systems in vitro are swelling, diffusion, and disintegration. In vitro degradation of chitosan-PMAA nanoparticles prepared by solution casing method occurred less rapidly as the degree 73% deacetylated showed slower biodegradation. Since the grade of chitosan used in the present study was of high molecular weight with a degree of deacetylation ≥85%, significant retardation of release of paclitaxel from nanoparticles is attributed to the polymer characteristics. In addition, diffusion of paclitaxel may have been hindered by increased tortuosity of polymer accompanied by a swelling mechanism.

Figure 4 shows the SEM of paclitaxel and chitosan-PMAA hydrogel film that synthesized by chemical reaction. This hydrogel is very sensitive to the temperature that due to the interaction of electron and sample. Scanning electron micrography images were obtained from a diluted solution of the paclitaxel particle. The white spots are paclitaxel nanoparticles. The SEM image shows the presence of paclitaxel spherical particles in hydrogel matrix, which are homogenenously distributed throughout the hydrogel, which is also confirmed from 1H NMR studies. As observed from SEM photomicrographs, the crystals of paclitaxel have a different appearance than recrystallized paclitaxel. These nanoparticles do not have clearly defined crystal morphological features in the SEM photomicrographs. Hence, it appears that the irregularly shaped particle are surface deposited with poloxamer, which gives them an appearance resembling that of coated particles. X-ray diffraction is also used to study the degree of crystallinity of pharmaceutical drugs and excipients. A lower 2Θ value indicates larger d-spacings, while an increase in the number of high-angle reflections indicates higher molecular state order. In addition, broadness of reflections, high noise, and low peak intensities are characteristic of a poorly crystalline material. A broad hump in the diffraction pattern of chitosan extending over a large range of 2Θ suggests.

CONCLUSIONS

The swelling and hydrolytic behavior of the hydrogel beads was dependent on the content of MAA groups and caused a decrease in gel swelling in SGF or an increase in gel swelling in SIF. Modified chitosan with different contents of MAA and CA by graft copolymerization reactions were carried out under microwave-radiation. The swelling of the hydrogels was dependent on the content of MAA groups and caused a decrease in gel swelling in SGF or an increase in gel swelling in SIF. Incorporation of MAA made the hydrogels pH-dependent and the transition between the swollen and the collapsed states occurred at high and low pH. The swelling ratios of the hydrogels beads increased at pH 7.4, but decreased at pH 1 with increasing incorporation of MAA.

REFERENCES

Bloembergen, N., & Pershan, P. S. (1967). Model catalysis of ammonia synthesis and iron–water interfaces – a sum frequency generation vibrational spectroscopic study of solid–gas interfaces and anion photoelectron spectroscopic study of selected anion clusters. *Physical Review,* 128(2), 606.

BrOndsted, H., & Kope˘cek, J. (1990). Hydrogels for site-specific oral delivery. In *Proceeding of the International Synposium on Controlled Release of Bioactive Materials,* vol. 17 (pp. 128–129).

Chiu, H. C., Hsiue, G. H., Lee, Y. P., & Huang, L. W. (1999). Synthesis and characterization of pH-sensitive dextran hydrogels as a potential colon-specific drug delivery system. *Journal of Biomaterials Science.* Polymer Edition, 10, 591–608.

Fanta, G. F., & Doane, W. N. (1986). Grafted starches. In O. B. Wurzburg (Ed.), *Modified Starches: Properties and Uses* (pp. 149–178). Boca Raton, FL: CRC.

Furda, I. (1983). Aminopolysaccharides – their potential as dietary fiber. In I. Furda (Ed.), *Unconventional Sources of Dietary Fiber, Physiological and In Vitro Functional Properties* (pp. 105–122). Washington, DC: American Chemical Society.

Giammona, G., Pitarresi, G., Cavallora, G., & Spadaro, G. (1999). New biodegradable hydrogels based on an acryloylated polyaspartamide crosslinked bygamma irradiation. *Journal of Biomedical Science.* Polymer Edition, 10, 969–987.

Jabbari, E., & Nozari, S. (2000). Swelling behavior of acrylic acid hydrogels prepared by γ-radiation crosslinking of polyacrylic acid in aqueous solution. *European Polymer Journal,* 36, 2685–2692.

Krogars, K., Heinamaki, J., Vesalahti, J., Marvola, M., Antikainen, O., & Yliruusi, J. (2000). Extrusion-spheronization of pH-sensitive polymeric matrix pellets for possible colonic drug delivery. *International Journal of Pharmacy,* 199, 187–194.

Mahfouz, A., Hamm, B., & Taupitz, M. (1997). Contrast agents for MR imaging of the liver: Clinical overview. *European Radiology,* 7, 507.

Peppas, N. A. (1987). Hydrogels in Medicine and Pharmacy. Boca Raton, FL: CRC Press. Puttpipatkhachorn, S., Nunthanid, J., & Yamamato, K. (2001). Drug physical state and drug–polymer interaction on drug release from chitosan matrix films. *Journal of Control Release,* 75, 143–153.

Ratner, B. D. (1989). Comprehensive polymer science – the synthesis, characterisation, reactions & applications of polymers. In S. K. Aggarwal (Ed.), *Comprehensive Polymer Science – The Synthesis, Characterization, Reactions and Applications of Polymers* (pp. 201–241). Oxford: Pergamon Press.

Saboktakin, M. R., Maharramov, A., & Ramazanov, M. A. (2007). Synthesis and characterization of aromatic polyether dendrimer/mesalamine (5-ASA) nanocomposite as drug carrier system. *Journal of American Science,* 3(4), 45.

Schmitz, S. A., Winterhalter, S., Schiffler, S., Gust, R., Wagner, S., Kresse, M., et al. (2000). Superparamagnetic iron oxide nanoparticles functionalized polymers. *Investigative Radiology,* 35, 460.

Thierry, B., Winnik, F. M., Mehri, Y., & Tabrizian, M. (2003). A new Y3Al5O12 phase produced by liquid-feed flame spray. *Journal of American Chemical Society,* 125, 7494.

Xu, H., & Li, T. (2005). The analysis of boundary functions of CMS reaction factors. *Journal of Nature and Science,* 3(2), 25–28.

In: Modern Nanochemistry
Eds: A. K. Haghi and G. E. Zaikov

ISBN: 978-1-61209-992-7
© 2011 Nova Science Publishers, Inc.

Chapter 2

SYNTHESIS OF LOCAL DELIVERY SYSTEMS BASED ON BIODEGRADABLE CHITOSAN BEADS

M. R. Saboktakin[1] and A. K. Haghi[2]

[1]Department of Nanotechnology, Baku State University, Azerbaijan
[2]University of Guilan, Iran

ABSTRACT

The main aim of this research is to design a new extended release gastroretentive multiparticulate delivery system by incorporation of the hydrogel beads made of chitosan. As the first part of a continued research on conversion of N-sulfonato-N, O-carboxymethylchitosan(NOCCS) to useful biopolymer-based materials, large numbers of carboxylic functional groups were introduced onto NOCCS by grafting with polymethacrylic acid (PMAA). The free radical graft copolymerization was carried out at 70 °C, bisacrylamide as a cross-linking agent and persulfate as an initiator. The equilibrium swelling studies were carried out in enzyme-free simulated gastric and intestinal fluids (SGF and SIF, respectively). Also, the satranidazole as a model drug was entrapped in nano-gels and in vitro release profiles were established separately in both enzyme-free SGF and SIF. The drug release was found to be faster in SIF. The drug release profiles indicate that the drug release depends on their degree of swelling and cross-linking.

INTRODUCTION

Natural polymers have potential pharmaceutical applications because of their low toxicity, biocompatibility, and excellent biodegradability. In recent years, biodegradable polymeric systems have gained importance for design of surgical devices, artificial organs, drug delivery systems with different routs of administration, carriers of immobilized enzymes and cells, biosensors, ocular inserts, and materials for orthopedic applications (BrOndsted & Kopecek, 1990). These polymers are classified as ei2ther synthetic (polyesters, polyamides, polyanhydrides) or natural (polyamino acids, polysaccharides) (Giammona et al., 1999; Krogars et al., 2000). Polysaccharide-based polymers represent a major class of biomaterials, which includes agarose, alginate, carageenan, dextran, and chitosan. Chitosan [_(1, 4)2-amino-2-d-glucose] is a cationic biopolymer produced by alkaline N-deacetylation of chitin, which is the main component of the shells of crab, shrimp, and krill (Chiu et al., 1999; Jabbari & Nozari, 2000).Chitosan is a functional linear polymer derived from chitin, the most abundant natural polysaccharide on the earth after cellulose, and it is not digested in the upper GI tract by human digestive enzymes (Fanta & Doane, 1986; Furda, 1983). Chitosan is a copolymer consisting of 2-amino-2-deoxy-d-glucose and 2-acetamido-2-deoxy-d-glucose units links with β-(1-4) bonds. It should be susceptible to glycosidic hydrolysis by microbial enzymes in the colon because it possesses glycosidic linkages similar to those of other enzymatically depolymerized polysaccharides. Among diverse approaches that are possible for modifying polysaccharides, grafting of synthetic

polymer is a convenient method for adding new properties to a polysaccharide with minimum loss of its initial properties (Peppas, 1987; Saboktakin et al., 2007). Graft copolymerization of vinyl monomers onto polysaccharides using free radical initiators, has attracted the interest of many scientists. Upto now, considerable works have been devoted to the grafting of vinyl monomers onto the substrates, especially Starch and cellulose (Honghua & Tiejing, 2005; Jabbari & Nozari, 2000). Existence of polar functionally groups as carboxylic acid need not only for bioadhesive properties but also for pH-sensitive properties of polymer (Ratner, 1989; Thierry et al., 2003). The increase of MAA content in the hydrogels provides more hydrogen bonds at low pH and more electrostatic repulsion at high pH. A part of our research program is chitosan modification to prepare materials with pH-sensitive properties for uses as drug delivery (Bloembergen & Pershan, 1967; Mahfouz et al., 1997; Schmitz et al., 2000). The free radical graft copolymerization polymethacrylic acid onto chitosan was carried out at 70 °C, bis-acrylamide as a cross-linking agent and persulfate as an initiator. Polymer bonded drug usually contain one solid drug bonded together in amatrix of a solid polymeric binder. They can be produced by polymerizing a monomer such as methacrylic acid (MAA), mixed with a particulate drug, by means of a chemical polymerization catalyst, such as AIBN or by means of high-energy radiation, such as X-ray or γ-rays. The modified hydrogel and satranidazole as a model drug were converted to nanoparticles by freeze-drying method. The equilibrium swelling studies and in vitro release profiles were carried out in enzyme-free simulated gastric and intestinal fluids [SGF (pH 1) and SIF (pH 7.4) respectively)]. The influences of different factors, such as content of MAA in the feed monomer and swelling were studied (Saboktakin et al., 2008).

MATERIALS AND METHODS

Derivatives of NOCCS, N-sulfonato-N, O-carboxymethylchitosan (NOCCS), containing sulfonato groups, SO_3 $-Na+$, were prepared by the reaction of sulfur trioxide–pyridine complex with NOCCS in alkaline medium at room temperature. Typically, 10 g of NOCCS (0.045 mol) dissolved in 0.60 L of water was treated with repeated additions of 12 g SO_3–pyridine. The SO_3–pyridine was slurried in 50–100mL of water and added dropwise, over 1 h. Both the NOCCS solution and the sulfur trioxide reagent slurry were maintained at a pH above 9 by the addition of sodium hydroxide (5 M). Following the last addition of the sulfating reagent, the sodium hydroxide solution was added until the pH stabilized (approx. 40 min.).

The pH of the mixture was adjusted to 9, the mixture was heated to 33 °C and held for 15 min. After filtration through a 110μm nylon screen, the filtered SNOCCS solution was poured into 6 L of 99% isopropanol. The resulting precipitate was collected and air dried overnight. The dried precipitate was dissolved in 0.45 L boiling water, solution was poured into dialysis sacks (M_W = 12, 000) and dialyzed for 3–4 days against dionized water. The contents of the sacks were lyophilized to yield the final product (6.2 g). N-sulfonato-N, O-carboxymethylchitosan with 1:1 molar ratios of methacrylic acid was polymerized at 60–70 °C in a thermostatic water bath, bis-acrylamide as a cross-linking agent (CA), using persulfate as an initiator ([I] = 0.02M) and water as the solvent (50 mL). The polymeric system was stirred by mechanical stirrer to sticky hydrogel and it was separated from medium without solvent addition. All experiments were carried out in Pyrex glass ampoules. After the specific time (48 h), the precipitated network polymer was collected and dried in vacuum. Copolymer (50 mg) and satranidazole (10 mg) were dispersed with stirring in 25mL deionised water. After approximately 180 min, the sample was sprayed into a liquid nitrogen bath cooled down to 77K, resulting in frozen droplets. These frozen droplets were then put into the chamber of the freeze-dryer. In the freeze drying process, the products are dried by a sublimation of the water component in an iced solution. The following procedure was used to assess the stability of satranidazole during the bead preparation process. The prepared beads were extracted twice with a solvent mixture of 1:1 acetonitrile and ethanol (v/v), the extract was evaporated, the residue was injected onto HPLC column. Stability-indicating chromatographic method was adopted for this purpose. The method consisted of a symmetry C18 column (254mm×4.6mm; 5_m) run using a mobile phase of composition methanol: water (70:30, v/v) at a flow rate of 0.5 mL/min, a Waters pump (600E), and eluants monitored with Water photodiode array detector (996 PDA) at 227 nm. A definite weight range of 10–15mg of bead was cut and placed in a 1.5mL capacity microcentrifuge tube containing 1mL of release medium of the following composition at 37 °C: phosphate buffered saline (140mM, pH 7.4) with 0.1% sodium azide and 0.1% Tween 80. At predetermined time points, 100_L of release medium was sampled with replacement to which 3mL of scintillation cocktail was added and vortexed before liquid scintillation counting. The cumulative amount of satranidazole released as a function of time was calculated. To study the molecular properties of satranidazole and Nsulfonato- N, O-carboxy-methylchitosan/ PMAA, the solid-state characterization of samples were done by the application of thermal, X-ray diffraction, and microscopy technique. During this study, the characteristics of satranidazole and Nsulfonato- N, O-carboxymethylchitosan/PMAA were compared with the beads to reveal any changes occurring as a result of bead preparation. Differential scanning calorimetry (DSC) studies were performed with a Mettler Toledo 821 thermal analyzer (Greifensee, Switzerland) calibrated with indium as standard. For thermogram acquisition, sample sizes of 1–5mgwere scanned with a heating rate of 5 °C/min over a temperature range of 25–300 °C. In order to check the reversibility of transition, samples were heated to a point just above the corresponding transition temperature, cooled to room temperature, and reheated up to 300 °C. Satranidazole samples and N-sulfonato-N, O-carboxymethylchitosan/ PMAA beads were viewed using a Philips XL-30 E SEM scanning electron microscope (SEM) at 30 kV (max.) for morphological examination. Powder samples of satranidazole and beads were mounted onto aluminium stubs using double-sided adhesive tape and then sputter coated with a thin layer of gold at 10 Torr vacuum before examination. The specimens were scanned with an electron beam of 1.2 kV acceleration potential, and images were collected in collected in secondary electron mode. Molecular arrangement of satranidazole and N-sulfonato-N, Ocarboxymethylchitosan/ PMAA in powder as well as in beads was compared by powder X-ray diffraction patterns acquired at room temperature on a Philips PW 1729 diffractometer (Eindhoven, Netherlands) using Cu-Kα radiation. The data were collected over an angular range from 3◦ to 50° 2Θ in continuous mode using a step size of 0.02° 2Θ and step time of 5 s.

RESULTS

In the present study, the beads were prepared by the classical method, which involves spreading a uniform layer of polymer dispersion followed by a drying step for removal of solvent system.

Since the methodology of bead preparation involved a heating step, it may have had a detrimental effect on the chemical stability of drug. Hence, the stability assessment of satranidazole impregnated in bead was done using stability-indicating method. For this purpose, satranidazole was extracted from bead and analyzed by HPLC. A single peak at 21.2m representing satranidazole (with no additional peaks) was detected in the chromatogram, suggesting that the molecule was stable during preparation of beads Satranidazole was extracted from different regions of Nsulfonato- N, O-carboxymethylchitosan/PMAA bead using acetonitrile: ETOH (1:1, v/v) solvent system. After normalization of satranidazole concentration on weight basis of bead, the variation in distribution of satranidazole in different regions of bead was <16%. The composition of the polymer defines its nature as a neutral or ionic network and furthermore, its hydrophilic/hydrophobic characteristics.

Ionic hydrogels, which could be cationic, containing basic functional groups or anionic, containing acidic functional groups, have been reported to be very sensitive to changes in the environmental pH. The swelling properties of the ionic hydrogels are unique due to the ionization of their pendent functional groups. Hydrogels containing basic functional groups is found increased swelling activity in acidic conditions and reduced in basic conditions. The pH-sensitive anionic hydrogels show low swelling activity in acidic medium and very high activity in basic medium. As shown in Figure 1, an increase in the content of MAA in the feed monomer mixtures resulted in less swelling in simulated gastric fluid but greater swelling in and simulated intestinal fluids.

This is because the increase of MAA content in the hydrogels provides more hydrogen bonds at low pH and more electrostatic repulsion at high pH. Figure 2 shows the scanning electron microscope (SEM) of graft NOCCS copolymer with polymethacrylic acid and nano-polymer bonded drug, respectively.

Nano- and micro-polymer bonded drugs (50 mg) were poured into 3mL of aqueous buffer solution (SGF: pH 1 or SIF: pH 7.4) (Figure 3). The mixture was introduced into a cellophane membrane dialysis bag. The bag was closed and transferred to a flask containing 20mL of the same solution maintained at 37 °C.

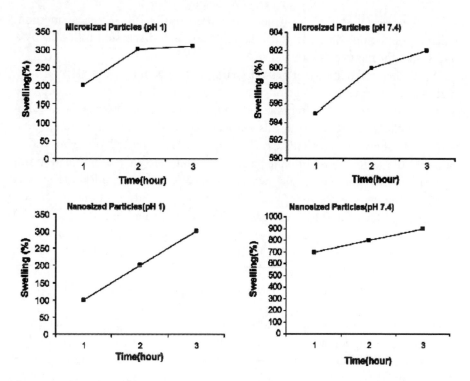

Figure 1. Time-dependent swelling of micro- and nano-carriers satranidazole drug model as a time at 37°C.

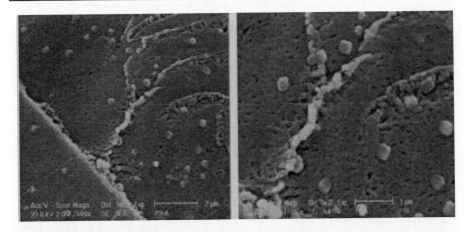

Figure 2. SEM of NOCCS-polumethacrylic acid hydrogel beads with satranidazole.

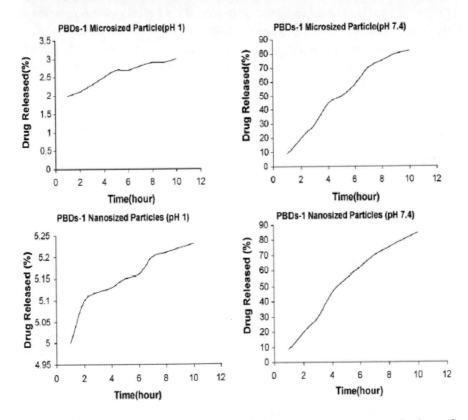

Figure 3. Release of satranidazole drug from micro- and nanocarriers as a fuctions of time at 37°C.

Table 1. DSC data and composition of copolymer

Polymer samples	Molar composition of monomers in the feed				Degree of substitution (DS)	T_g (°C)
	NOCCS (g)	MAA (g)	CA (g)	IN (g)		
P-1	1	3	0.05	0.05	0.52	135
P-2	1	2	0.05	0.05	0.49	142

The external solution was continuously stirred, and 3mL samples were removed at selected intervals. The removed volume was replaced with SGF or SIF Figure 3.

The triplicate samples were analyzed by UV spectrophotometer, and the quantity of satranidazole were determined using a standard calibration curve obtained under the same conditions. Nano- and micro-polymer bonded drugs (50 mg) were poured into 3mL of aqueous buffer solution (SGF: pH 1 or SIF: pH 7.4) (Figure 3). The mixture was introduced into a cellophane membrane dialysis bag. The bag was closed and transferred to a flask containing 20mL of the same solution maintained at 37 °C. The external solution was continuously stirred, and 3mL samples were removed at selected intervals. The removed volume was replaced with SGF or SIF Figure 4.

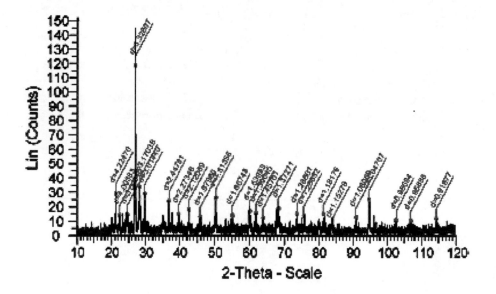

Figure 4. XRD pattern of NOCCS hydrogel beads.

The triplicate samples were analyzed by UV spectrophotometer, and the quantity of satranidazole were determined using a standard calibration curve obtained under the same conditions. It appears that the degree of swelling depends on their particle size, a decrease in the molecular size of carriers increased the swelling rate. The thermal behavior of a polymer is important for controlling the release rate in order to have a suitable drug dosage form. The glass transition temperature (Tg) was determined from the DSC thermograms. The values are given in Table 1.

The higher Tg values probably related to the introduction of cross-links, which would decrease the flexibility of the chains and the ability of the chains to undergo segmental motion, which would increase the Tg values. On the other hand, the introduction of a strongly polar carboxylic acid group can increase the Tg value because of the formation of internal hydrogen bonds between the polymer chains. X-ray diffraction is a proven tool to study crystal lattice arrangements and it yields very useful information on degree of crystallinity. The X-ray diffraction patterns of hydrogel with satranidazole have several high-angle bead at the following 2Θ values: $21°$, $22°$, $25°$, $26.5°$, $28°$, and $32°$. The $26.5°$ 2Θ peak had the highest intensity, and the hump in the baseline occurred from $7°$ to $45°$ 2Θ, as observed for chitosan bead.

DISCUSSION

In order to develop a local delivery system for satranidazole, a biodegradable N-sulfonato-N, O-carboxymethylchitosan–PMAA bead was attempted, which to our knowledge in the first effort of its kind. The potential formulation problems were anticipated since chitosan is only soluble in aqueous acidic solutions, whereas satranidazole, being a hydrophobic drug, is insoluble under similar conditions. In early stages of formula optimization studies, the satranidazole was incorporated into bead. The primary mechanisms for release of drugs from matrix systems in vitro are swelling, diffusion, and disintegration. In vitro degradation of chitosan beads were prepared by solution casing method occurred less rapidly as the degree 73% deacetylated showed slower biodegradation. Since the grade of chitosan used in the present study was of high molecular weight with a degree of deacetylation $\geq 85\%$, significant retardation of release of satranidazole from bead is attributed to the polymer characteristics. In addition, diffusion of satranidazole may have been hindered by increased tortuosity of polymer accompanied by a swelling mechanism. As observed from SEM photomicrographs, the crystals of satranidazole have a different appearance than recrystallized satranidazole. These nanoparticles do not have clearly defined crystal morphological features in the SEM photomicrographs. Hence, it appears that the irregularly shaped particle are surface deposited with poloxamer, which gives them an appearance resembling that of coated particles. X-ray diffraction technique is also used to study the degree of crystallinity of pharmaceutical drugs and excipients. A lower 2Θ value indicates larger d-spacings, while an increase in the number of high-angle reflections indicates higher molecular state order. In addition, broadness of reflections, high noise, and low peak intensities are characteristic of a poorly crystalline material. A broad hump in the diffraction pattern of chitosan extending over a large range of 2Θ suggests that chitosan is present in amorphous state in the bead.

CONCLUSIONS

The swelling and hydrolytic behavior of the hydrogels beads were dependent on the content of MAA groups and caused a decrease in gel swelling in SGF or an increase in gel swelling in SIF.

Modified chitosan with different contents of MAA and CA by graft copolymerization reactions were carried out under microwaveradiation. The swelling of the hydrogels beads was dependent on the content of MAA groups and caused a decrease in gel swelling in SGF or an increase in gel swelling in SIF. Incorporation of MAA made the hydrogels beads pH-dependent and the transition between the swollen and the collapsed states occurred at high and low pH. The swelling ratios of the hydrogels beads increased at pH 7.4, but decreased at pH 1 with increasing incorporation of MAA.

REFERENCES

Bloembergen, N., & Pershan, P. S. (1967). Model catalysis of ammonia synthesis of iron–water interfaces – a sum frequency generation vibrational spectroscopic study of solid–gas interfaces and anion photoelectron spectroscopic study of selected anion clusters. *Physical Review,* 128(2), 606.

BrOndsted, H., &Kope˘cek, J. (1990). Hydrogels for site-specific oral delivery. *Proceedings of the International Synposium on Controlled Release of Bioactive Materials,* 17, 128–129.

Chiu, H. C., Hsiue, G. H., Lee, Y. P., & Huang, L. W. (1999). Synthesis and characterization of pH-sensitive dextran hydrogels as a potential colon-specific drug delivery system. *Journal of Biomaterials Science,* Polymer Edition, 10, 591–608.

Fanta, G. F., & Doane, W. N. (1986). Grafted starches. In O. B. Wurzburg (Ed.), Modified starches: properties and uses (pp. 149–178). *Boca Raton (FL): CRC.*

Furda, I. (1983). Aminopolysaccharides – their potential as dietary fiber. In I. Furda (Ed.), *Unconventional sources of dietary fiber, physiological and in vitro functional properties* (pp. 105–122). Washington, DC: American Chemical Society.

Giammona, G., Pitarresi, G., Cavallora, G., & Spadaro, G. (1999). New biodegradable hydrogels based on an acryloylated polyaspartamide crosslinked by gamma irradiation. *Journal of Biomedical Science,* Polymer Edition, 10, 969–987.

Honghua, Xu., & Tiejing, Li. (2005). The analysis of boundary functions of CMS reaction factors. *Journal of Nature and Science,* 3(2), 25–28.

Jabbari, E., & Nozari, S. (2000). Swelling behavior of acrylic acid hydrogels prepared by γ-radiation crosslinking of polyacrylic acid in aqueous solution. *European Polymer Journal,* 36, 2685–2692.

Krogars, K., Heinamaki, J., Vesalahti, J., Marvola, M., Antikainen, O., & Yliruusi, J. (2000). Extrusion-spheronization of pH-sensitive polymeric matrix pellets for possible colonic drug delivery. *International Journal of Pharmaceutics,* 187–194.

Mahfouz, A., Hamm, B., & Taupitz, M. (1997). Contrast agents for MR imaging of the liver: clinical overview. *European Radiology,* 7, 507.

Peppas, N. A. (1987). Hydrogels in medicine and pharmacy. Boca Raton, FL: CRC Press.

Ratner, B. D. (1989). S. K. Aggarwal (Ed.), *Comprehensive polymer science – the synthesis, characterisation, reactions & applications of polymers* (pp. 201–247). Oxford: Pergamon Press.

Saboktakin, M. R., Maharramov, A., &Ramazanov, M. A. (2007). Synthesis and characterization of aromatic polyether dendrimer/mesalamine(5-ASA) nanocomposite as drug carrier system. *Journal of American Science,* 3(4), 45.

Saboktakin, M. R., Maharramov, A., & Ramazanov, M. A. (2008). Poly (amidoamine) (PAMAM)/CMS dendritic nanocomposite for controlled drug delivery. *Journal of American Science,* 4(1), 48.

Schmitz, S. A., Winterhalter, S., Schiffler, S., Gust, R., Wagner, S., Kresse, M., Coupland, S. E., Semmler, W., & Wolf, K. J. (2000). Superparamagnetic iron oxide nanoparticles functionalized polymers. *Investigative Radiology,* 35, 460.

Thierry, B., Winnik, F. M., Mehri, Y., & Tabrizian, M. (2003). A new Y3Al5O12 phase produced by liquid-feed flame spray. *Journal of American Chemical Society,* 125, 7494.

In: Modern Nanochemistry
Eds: A. K. Haghi and G. E. Zaikov

ISBN: 978-1-61209-992-7
© 2011 Nova Science Publishers, Inc.

Chapter 3

MRI DETECTABLE COLON DRUG DELIVERY SYSTEMS BASED ON SUPERPARAMAGNETIC CHITOSAN-DEXTRAN SULFATE HYDROGELS

M. R. Saboktakin[1] and A. K. Haghi[2]

[1]Department of Nanotechnology, Baku State University, Azerbaijan
[2]University of Guilan, Iran

ABSTRACT

The purpose of this study was to examine chitosan (CS)–dextran sulfate (DS) nanoparticles coated iron oxide as drug carriers detectable using magnetic resonance imaging (MRI) technique. The 5-aminosalicylic acid (5-ASA) was chosen as model drug molecule. CS–DS hydrogels were formulated by a complex coacervation process under mild conditions. The influence of process variables, including the two ionic polymers, on particle size, and hydrogel entrapment of 5-ASA was studied. The in vitro release of 5-ASA were also evaluated, and the integrity of 5-ASA in the release fraction was assessed using sodium dodecyl sulfate-polyacrylamide gel electrophoresis. The release of 5-ASA from hydrogel was based on the ion-exchange mechanism. The CS–DS hydrogel developed based on the modulation of ratio show promise as a system for controlled delivery of drug detectable using magnetic resonance imaging (MRI) technique.

INTRODUCTION

Targeting of drugs specifically to the colon is advantageous in the treatment of diseases such as amoebiasis, Crohn's disease, ulcerative colitis, and colorectal cancer. In addition, it has shown great potential in the oral delivery of therapeutic peptides and proteins, which are unstable in the upper part of the gastrointestinal (GI) tract. The colonic region is recognized as having less diversity and intensity of enzymatic activities than stomach and small intestine (Davis, 1990). Various strategies are available for targeting drug release selectively to the colon (Chourasia & Jain, 2003). The designing of prodrugs is based on the concept of preventing the release of drugs in the stomach and small intestine and drug release is triggered by the use of specific property at the target site such as altered pH or high activity of certain enzymes in comparison to nontarget tissues (Davaran, Hanaee, & Khosravi, 1999; Schacht et al., 1996). Since it is known that azo function can be reduced in the colon (Chung, Stevens, & Cerniglia, 1992), many novel polymers containing azo groups either in the polymeric backbone (Yamaoka, Makita, Sasatani, Kim, & Kimura, 2000) or in the cross-links (Shantha, Ravichandran, & Rao, 1995; Van den Mooter, Samyn, &Kinget, 1992) have been synthesized. To promote further selective degradation in the vicinity of the colonic environment, delivery systems have been designed that contain both pH-sensitive acidic monomers and degradable azo aromatic crosslinks (Ghandehari, Kopeckova, &Kopecek, 1997; Kakoulides, Smart, & Tsibouklis, 1998). Chitosan is a functional linear polymer derived from chitin, the most abundant natural polysaccharide on the earth after cellulose, and it is not digested in the upper GI tract by human digestive enzymes (Furda, 1983; Ormrod, Holmes, & Miller, 1998). Chtosan is a copolymer consisting of 2-amino-2-deoxyd- glucose and 2-acetamido-2-deoxy-d-glucose units links with _-(1–4) bonds. It should be susceptible to glycosidic hydrolysis by microbial enzymes in the colon because it possesses glycosidic linkages similar to those of other enzymatically depolymerized polysaccharides. The polysaccharide, on reaching the colon, undergoes assimilation by microorganisms or degradation by enzymes or break down of the polymer back bone leading to a subsequent reduction in molecular weight and thereby loss of mechanical strength and is unable to hold the drug any longer (Yamamoto, Tozaki, Okada, & Fujita, 2000). Chitosan has drawn attention for its potential to achieve site-specific delivery to the colon. Chitosan, a natural linear polyamine with a high ratio of glucosamine to acetyl-glucosamine units, is a weak base and carriers a positive charge. Its solubility is pH-dependent, and it reacts readily with negatively charged surfaces and materials, including polymers and DNA. Ionic gelation, complex coacervation, emulsion crosslinking, and spray-drying are methods commonly used for the preparation of chitosan nanoparticles (Agnihotri, Mallikarjuna, & Aminabhavi, 2004). Among those methods, ionic gelation and complex coacervation are mild processes occurring in a pure aqueous environment and are ideal for maintaining the in-process stability of drugs. Ionic gelation and complex coacervation are very similar except that the former involves the gelation of chitosan using an electrolyte such as tripolyphosphate (TPP) (Calvo, Remunan- Lopez, Vila-Jato, & Alonso, 1997), whereas the latter employs an oppositely charged ionic polymer such as alginate (De & Robinson, 2003). A new type of chitosan (CS) hydrogel using dextran sulfate (DS) as a polyanionic polymer was developed to achieve complex coacervation for the incorporation and controlled release of an anti-angiogenesis hexapeptide (Chen, Mohanraj, & Parkin, 2003), this was the first report describing the use of DS to formulate CS based hydrogels. Although there have been investigations of how the properties of CS and formulation variables such as CS molecular weight (M_W), concentrations of CS and 5-ASA, and formulation pH affect the formation and encapsulation capability of hydrogels (Ma, Yeoh, & Lim, 2002; Tiyaboonchai, Woiszwillo, Sims, & Middaugh, 2003), to our knowledge, no attempts have been made to study how the ratio of CS to the oppositely charged polymer influences the formulation and properties of hydrogel. In this study, 5- aminosalicylic acid (5-ASA) was chosen as amodel drug (Pan et al., 2002).

MATERIALS AND METHODS

Chitosan (CS) and dextran sulfate (DS) hydrogels were prepared using the method reported by Zhang et al., with slight modifications. Chitosan–dextran sulfate (at the charge ratio (N:P) of 1.12) and 35mg $FeCl_3 \cdot 6H_2O$ were solved in 4mL H_2O and nitrogen was flushed for 1.5 h. 14mg $FeCl_2 \cdot 4H_2O$ was added, followed by 100μL aqueous ammonia in two portions while the mixture was kept under nitrogen. The solution turned black and was heated to 80 °C for 100 min. After the mixture cooled to room temperature, the ammonia was removed by flushing the solution with nitrogen over 10 min. Then, the solid mixture was dissolved in 10mL acetic acid (1%, w/v) and 5-aminosalicylic acid (5-ASA) (5–20 mg) was dissolved separately in dimethyl formamide (DMF). Then, the magnetic chitosan/dextran sulfate (2.5 mL) and the drug solution

(7.5 mL) were mixed together to obtain 10mL of chitosan/dextran sulfate drug solution. The chitosan/dextran sulfate drug solution was added dropwise (using a disposable syringe with a 22-gauge needle) into 40mL of sodium-saturated Tris–HCl buffer solution containing glutaraldehyde-saturated toluene (GST) in different concentrations (1–3 mL) under magnetic stirring (~200 rpm) at room temperature. The hydrogel suspension was formed spontaneously.

The mixture was stirred for a further 15 min. The hydrogel was separated after 1 h of curing time and subsequently decanted, washed twice with 3mL of 0.05MTris–HCl buffer, and the hydrogel was dried in vacuum oven at 40 ∘C. Surface and shape characteristics of chitosan–dextran sulfate hydrogels were evaluated by means of a scanning electron microscope (FEI-Qunta-200 SEM, FEI Company, Hillsboro, OR). The samples for SEM were prepared by lightly sprinkling the hydrogel on a bouble adhesive tape, which stuck to an aluminum stub. The stubs were than coated with gold to a thickness of ~300Å using a sputter coater and viewed under the scanning electron microscope. After drying at 37 ∘C for 48 h, the mean diameter of the dried hydrogel was determined by a sieving method using USP standard sieves. Observations are recorded. Encapsulation efficiency (EE) is the amount of added drug (%) that is encapsulated in the formulation of the hydrogel. The EE of drug from hydrogel can be calculated in terms of the ratio of drug in the final formulation to the amount of added drug. An accurately weighed amount (100 mg) of the formulation of hydrogel was dispersed in 100mL of Tris–HCl buffer. The sample was ultrasonicated for three consecutive periods of 5min each, with a resting period of 5min each. It was left to centrifuged at 3000rpm for 15 min. The concentration of 5-aminosalicylic acid (5-ASA) in the decanted Tris–HCl buffer and two washing solutions was determined by measuring the absorbance at 235nm using a GBS Cintra 10-UV–visible spectrophotometer (Shimadzu, Japan). The determinations were made in triplicate, and results were averaged (Table 1). Chitosan–dextran sulfate hydrogels (100 mg) were placed in phosphate buffered saline (PBS) (pH 7.4) and allowed to swell upto a constant weight. The hydrogels were removed, blotted with filter paper, and changes in weight were measured and recorded in Table 1.

The degree of swelling (%) was then calculated from the formula:

$$Wg - Wo/ Wo$$

where, Wo is the initial weight of hydrogel andWg is the weight of hydrogel at equilibrium swelling in the medium. In vitro drug release studies were performed according to extraction technique using USP dissolution test apparatus. The dissolution studies were performed in 900mL of dissolution medium, which was stirred at 100rpm at 37±0.1 °C.

Table 1. Composition and characterization of different chitosan-dextran sulfate hydrogel

Variables	Values	Degree of swelling	Average particle size (mm)	Encapsulation efficiency (%)
5-ASA	0	1.36 ± 0.05	1.20 ± 0.08	–
	5	1.24 ± 0.20	1.68 ± 0.08	74.8 ± 2.34
	10	1.00 ± 0.01	1.70 ± 0.06	75.2 ± 1.98
	15	0.95 ± 0.09	1.92 ± 0.04	80.1 ± 2.54
	20	0.68 ± 0.12	1.20 ± 0.10	85.6 ± 1.63
CS–DS hydrogel (%w/w)	1	1.10 ± 1.00	1.38 ± 0.11	80.8 ± 3.24
	2	1.20 ± 1.21	1.75 ± 0.09	82.3 ± 1.25
	3	1.32 ± 1.35	1.90 ± 0.05	86.7 ± 2.47
	4	1.50 ± 1.40	1.87 ± 0.02	88.7 ± 2.35
Time (h)	10 min	0.50 ± 0.52	1.90 ± 0.12	70.3 ± 1.47
	20 min	0.53 ± 0.14	1.87 ± 0.21	86.4 ± 1.98
	40 min	0.85 ± 0.05	1.70 ± 0.32	76.7 ± 2.45
Drying (°C)	Lyophilization	0.86 ± 0.50	1.32 ± 0.12	69.8 ± 4.35
	45	0.71 ± 0.72	1.00 ± 0.64	65.4 ± 3.98

The scheme of using the simulated fluids at different pH was as follows:

- First hour: simulated gastric fluid of pH 1.2
- Second to third hour: mixture of simulated gastric and intestinal fluid of pH 4.5
- Fourth to Fifth hour: simulated intestinal fluid of pH 7.4
- Sixth to eighth hour: simulated colonic fluid of pH 7.0

In vitro drug release studies were performed as per the scheme in different simulated fluids. Simulation of GI transit conditions was achieved by using different dissolution media. Simulated gastric fluid (SGF) pH 1.2 consisted of NaCl (0.2 g), HCl (7 mL), and pesin (3.2 g); pH was adjusted to 1.2±0.5. Simulated intestinal fluid (SIF) pH 7.4 consisted of KH2PO4 (6.8 g), 0.2N NaOH (190 mL), and pancreatin (10.0 g); pH was adjusted to 7.4±0.1. SIF pH 4.5 was prepared by mixing SGF pH 1.2 and SIF pH 7.4 in a ratio of 36:61. The experiment was performed with a continuous supply of carbon dioxide into dissolution media. Aliquots of samples were withdrawn periodically and replaced with an equal amount of fresh dissolution media bubbled with carbon dioxide. The volume was made upto 10mL and centrifuged. The supernatant was filtered through Whatman filter paper (Dawsonville, GA), and drug content was determined spectrophotometrically at 235nm (UV 1601,

Shimadzu, Japan). Experimental data have been represented as the mean with standard deviation (SD) of different independent determinations.

The significance of differences was evaluated by analysis of variance (ANOVA). Differences were considered statistically significant at $P < 0.005$. Conductivity of hydrogel dispersion, 0.1% CS, 0.1%DS with iron oxide nanoparticles in the deionized water was determined using a conductivity meter (Systronics 307) with a conductivity range from 0.1 to 200m_ at room temperature. All samples were prepared in deionized water.

RESULTS AND DISCUSSION

These hydrogels had good spherical geometry. It is obvious that the surface of the hydrogel shrank and a densely cross-linked gel structure was formed. This may explain the greater retardationof drug release from matrices of higher cross linker content. The average drug entrapment was found to be $81.21\pm1.86\%$ in the hydrogel. Results of drug content and EE demonstrated that drug content increased from 10.23 ± 0.50 mg/100mg to 26.24 ± 0.43 mg/100mg of hydrogel with increasing theamountof drug from 5 to 20%(w/w).

Figure 1. SEM of Chitosan- dextran sulfate hydrogel.

No signification increase in drug content was observed on further increasing the amount of drug, i.e., above 15% (w/w), which could be due to the limited solubility of the drug in DMF and that is endorsed from the presence of drug particles on the surface of the hydrogel prepared with 20% of drug concentration. The percent EE was increased upto 86.32±0.20% with increasing polymer concentrations to 4%. Concentration of the cross-linking agent exhibited no significant effect on percent EE. FESEM images of superparamagnetic CS–DS hydrogel (Figure 1) show that hydrogel have a solid and near consistent structure.

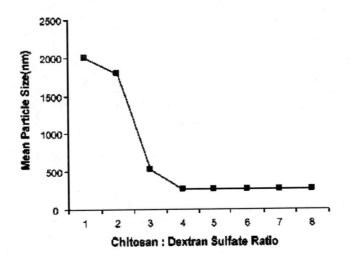

Figure 2. Influence of the chitosan: dextran sulfate ratio on the particle size.

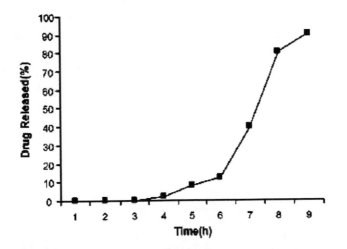

Figure 3. the in vitro release profile of 5-aminosalicylic acid from chitosan-dextran sulfate hydrogel in various simulated gastrointestinal fluids(n=3).

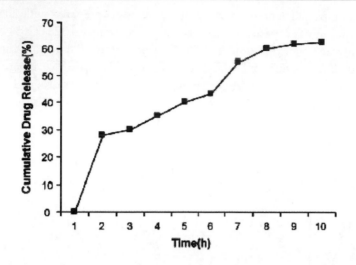

Figure 4. In vitro drug release from chitosan-dextran sulfate hydrogel in SCF(pH 7.0).

Furthermore, the incorporation of 5-ASA into the hydrogel produced a smooth surface and compact structure. The particle size observed in FESEM is smaller than that measured by the Zetasizer. This is because dried hydrogel was used in FESEM, whereas particles in the liquid dispersionwasanalyzed by the Zetasizer. CS–DS hydrogel are hydrophilic and would be expected to swell in water, thus producing a large hydrodynamic size when measured by the Zetasizer (Figure 2). 5-aminosalicylic acid (5-ASA)-loaded hydrogel was obtained spontaneously upon the mixing of the DS aqueous solution (0.1%, w/v) with the CS solution (0.1%, w/v) under magnetic stirring, with 5-ASA dissolved in CS–DS solution. The incorporation of 5-ASA into the CS–DS hydrogel resulted in a sharp increase in the particle size of the nanoparticle dispersion. The significant increases in particle size give a good induction of the incorporation of 5-ASA into CS–DS hydrogel. A study was undertaken to investigate the effect of the order of 5-ASA mixing with CS and DS.

Table 2. Conductivity of dispersion of hydrogel in the different pH

Weight ratio of CS:DS	Charge ratio (N:P) of the hydrogel	pH of hydrogel	Conductivity of dispersion (in mʊ)	Size (in nm)
5:4.5	1.01	4.0	3.72	1892
5:5	1.12	3.9	3.70	350
5:8.5	1.90	4.1	3.78	240

The data obtained show that the order of 5-ASA mixing had no effect on the size, entrapment efficiency, and yield of 5-ASA-loaded hydrogel. The effect of drug concentration, chitosan–dextran sulfate concentration, GST concentration, and cross-linking time on in vitro drug release was also observed. In vitro drug release after 5 h was 86.3±4.0% in the case of hydrogel having 15% drug, while it was 88.2±3.1% for hydrogel with 20% drug. The effect of chitosan–dextran sulfate on the release of drug was found to be meager. It is also observed that the amount of drug released from hydrogel decreased on increasing cross-linking time. These properties are probably explained by the promotion of cross-links between chitosan–dextran sulfate chains and GST. Freeze-drying of the samples resulted in larger and more porous hydrogel compared with air-dried hydrogel. Freeze-drying had the advantage of avoiding drug extraction by immediately freezing and removing the water present within the hydrogel and a burst effect during the dissolution study (Figure 3). Conventional dissolution testing is less likely to accurately predict in vivo performance of colon delivery systems triggered by bacteria residing in the colon (because aspects of the colon's environment, i.e., scarcity of fluid, reduced motility, and presence of microflora, cannot be simulated in conventional dissolution methods).

Hence, release studies were performed in an alternate release medium. The release profiles of 5-ASA-loaded hydrogel was evaluated in water and a phosphate buffer, which was either in a different ionic strength or with saline, to study theunderprinning mechanisms for drug release. The greatest release for 5-ASA–loaded hydrogel occurred in the release media of a high ionic strength. In contrast, a significantly small portion of 5-ASA was released in water over the release study period. The burst release was observed with hydrogel, and it may have arisen from the desorption of loosely attached 5-ASA from the surface of the matrix polymers (Figure 4). The opposite charges of CS–DS hydrogel were responsible for the formation of nanoparticles. The charge ration between the negatively charged sulfate groups (N) in DS and the positively charged amine groups (P) in CS was determined. Under the experimental conditions (pH 3–4), DS carries ~74 sulfate groups per mole, equivalent to 5.78×10^{-3} negatively charged groups per gram of DS; CS has~2073 amino groups per gram of CS. Table 2 shows conductivity of dispersion of hydrogel.

CONCLUSIONS

Results of release studies indicate that superparamagnetic chitosan–dextran sulfate hydrogel offer a high degree of protection from premature drug release in simulated upper conditions. These hydrogels deliver most of the drug load in the colon, an environment rich in bacterial enzymes that degrade the chitosan–dextran sulfate and allow drug release to occur at the desired site. Thus, spherical superparamagnetic hydrogel is a potential system for colon delivery of 5-ASA. Also, this hydrogel can be detect by magnetic resonance imagining (MRI) technique.

REFERENCES

Agnihotri, S. A., Mallikarjuna, N. N., & Aminabhavi, T. M. (2004). Recent advances on chitosan-based micro- and nanoparticles in drug delivery. *Journal of Controlled Release,* 100, 5–28.

Calvo, P., Remunan-Lopez, C., Vila-Jato, J. L., & Alonso, M. J. (1997). Novel hydrophilic chitosan-polyethylene oxide nanoparticles as protein carriers. *Journal of Applied Polymer Science,* 63, 125–132.

Chen, Y., Mohanraj, V., & Parkin, J. (2003). Chitosan–dextransulfate nanoparticles for delivery of an anti-angiogenesis peptide. International *Journal of Peptide Research and Therapeutics*, 10, 621–629.

Chourasia, M. K., & Jain, S. K. (2003). Pharmaceutical approaches to colon targeted drug delivery systems. *Journal of Pharmaceutical Sciences*, 6, 33–66.

Chung, K. T., Stevens, S. E., & Cerniglia, C. E. (1992). The reduction of azo dyes by the intestinal microflora. *Critical Reviews of Microbiology*, 18, 175–190.

Davaran, S., Hanaee, J., & Khosravi, A. (1999). Release of 5-aminosalicylic acid from acrylic type polymeric prodrugs designed for colon-specific drug delivery. *Journal of Controlled Release,* 58, 279–287.

Davis, S. S. (1990). Assessment of gastrointestinal transit and drug absorption. In L. F. Prescott, &W. S. Nimmo (Eds.), *Novel drug delivery and its therapeutic application* (pp. 89–101). Chichester, UK: Wiley.

De, S., & Robinson, D. (2003). Polymer relationships during preparation of chitosan - alginate and poly-l-lysine-alginate nanospheres. *Journal of Controlled Release,* 89, 101–112.

Furda, I. (1983). Aminopolysaccharides—their potential as dietary fiber. In I. Furda (Ed.), *Unconventional sources of dietary fiber, physiological and in vitro functional properties* (pp. 105–122). Washington, DC: American Chemical Society.

Ghandehari, H., Kopeckova, P., & Kopecek, J. (1997). In vitro degradation of pH sensitive hydrogels containing aromatic azo bonds. *Biomaterials,* 18, 861–872.

Kakoulides, E. P., Smart, J. D., & Tsibouklis, J. (1998). Azo crosslinked poly(acrylic acid) for colonic delivery and adhesion specificity synthesis and characterization. *Journal of Controlled Release,* 52, 291–300.

Ma, Z., Yeoh, H. H., & Lim, L. Y. (2002). Formulation pH modulates the interaction of insulin with chitosan nanoparticles. *Journal of Pharmaceutical Sciences,* 91, 1396–1404.

Ormrod, D. J., Holmes, C. C., & Miller, T. E. (1998). Dietary Chitosan inhibits hypercholesterolaemia and atherogenesis in the apolipoprotein E-deficient mouse model of atherosclerosis. *Atherosclerosis,* 138, 329–334.

Pan, Y., Li, j., & Zhao, H. (2002). Bioadhesive polysaccharide in protein delivery system: Chitosan nanoparticles improve the intestinal absorption of insulin in vivo. *International Journal of Pharmaceutics,* 24, 139–147.

Schacht, E., Gevaert, A., & Kenawy, E. R. (1996). Polymers for colon specific drug delivery. *Journal of Controlled Release,* 58, 327–338.

Shantha, K. L., Ravichandran, P., & Rao, K. P. (1995). Azo polymeric hydrogels for colon targeted drug delivery. *Biomaterials,* 16, 1313–1318.

Tiyaboonchai, W., Woiszwillo, J., Sims, R. C., &Middaugh, C. R. (2003). Insulin containing polyethylenimine-dextran sulfate nanoparticles. *Journal of Pharmaceutical Sciences,* 255, 139–151.

Van den Mooter, G., Samyn, C., & Kinget, R. (1992). Azo polymers for colon-specific drug delivery. *International Journal of Pharmaceutics,* 87, 37–46.

Yamamoto, A., Tozaki, H., Okada, N., & Fujita, T. (2000). Colon specific delivery of peptide drugs and anti-inflammatory drugs using chitosan capsules. *STP Pharma Science,* 10, 23–43.

Yamaoka, T., Makita, Y., Sasatani, H., Kim, S. I., & Kimura, Y. (2000). Linear type azocontaining polyurethane as drug-coating material for colon-specific delivery: Its properties degradation behavior and utilization for drug formulation. *Journal of Controlled Release,* 66, 187–197.

In: Modern Nanochemistry
Eds: A. K. Haghi and G. E. Zaikov

ISBN: 978-1-61209-992-7
© 2011 Nova Science Publishers, Inc.

Chapter 4

SYNTHESIS OF MRI DETECTABLE DRUG DELIVERY SYSTEMS BASED ON AMINODEXTRAN-COATED IRON OXIDE NANOPARTICLES

M. R. Saboktakin[1] and A. K. Haghi[2]

[1]Department of Nanotechnology, Baku State University, Azerbaijan
[2]University of Guilan, Iran

ABSTRACT

A new synthetic macromolecule, aminodextran-coated iron oxide nanoparticles, was synthesized as drug carrier detectable using magnetic resonance imaging (MRI) technique. The synthesis process starts with a 2-step reaction that attaches a high density of amino groups to a dextran backbone. These macromolecules were coated with magnetic iron oxide molecules by a chemical reaction that can carry several molecules such as drug and peptides. The aminodextran-coated iron oxide nanoparticles thus synthesized have been characterized by Fourier infrared (FT-IR) spectroscopy. Also, the morphology of this synthetic macromolecule was studied by scanning electron microscopy.

INTRODUCTION

Nanomedicine is an emerging field that uses nanoparticles to facilitate the diagnosis and treatment of diseases. Notable early successes in the clinic include the use of magnetic nanoparticles as a contrast agent in MRI and nanoparticle-based treatment systems (Vera, Buonocore, Wisner, Katzberg, & Stadalnik, 1995). The first generation of nanoparticles used in tumor treatments rely on leakiness of tumor vessels for preferential accumulation in tumor; however, this enhanced permeability and retention is not a constant feature of tumor vessels and even when present, still leaves the nanoparticles to negotiate the high interstitial fluid pressure in tumors (Morton, Wen, & Wong, 1992; Vera, Wisner, & Stadalnik, 1997). An attractive alternative would be to target nanoparticles to specific molecular receptors in the blood vessels because they are readily available for binding from the blood stream and because tumor vessels express a wealth of molecules that are not significantly expressed in the vessels of normal tissues (Gershenwald, Tseng, & Thomson, 1998; Glass, Messina, & Cruse, 1996).

Specific targeting of nanoparticles to tumors has been accomplished in various experimental systems (Giuliano, Kirgan, Guenther, & Morton, 1994), but the efficiency of delivery is generally low. An amplified homing is an important mechanism that ensures sufficient platelet accumulation at the sites of vascular injury (Chipowsky& Lee, 1973; Lee, Stowell, & Krantz, 1976). Amplified homing involves target binding, activation, platelet–platelet binding, and formation of a blood clot. Nanoparticle-based diagnostics and therapeutics hold great promise because multiple functions can be built into the particles (Qu, Wang, Zhu, Rusckowski, & Hnatowich, 2001; Shirakami et al., 1987). One such function is an ability to home to specific sites in the body. A research describe the magnetic particles new application not only home to tumors, but also amplify own homing (Porter, 1997; Tomalia, Baker, & Deward, 1985). The system is based on a magnetic drug–dextran conjugate that carries anti-tumor drug molecules (Nugent & Jain, 1984; Thoren, 1978). Iron oxide nanoparticles coated with this dextran–drug conjugate thereby producing new binding sites for more particles (Korosy & Barczai-Martos, 1950; Krejcarek & Tucker, 1977). A nanoparticle delivery system has been designed in which the particles amplify their own homing in a manner that resembles platelet (de Belder, 1996).

MATERIALS AND METHODS

PM70 dextran (10 g) in 75 mL deionized water was prepared at 50 °C and pH 11 in the presence of 2.5 g sodium hydroxide and 0.2 g sodium borohydride. The pH was maintained by dropwise addition of 2.5N NaOH and 2 mL allyl bromide. The solution was neutralized with acetic acid (2.5 mol/L) and the reaction mixture was placed in a 5 °C refrigerator for 2 h. After the top organic layer was decanted and 100 mL deionized water was added, the resulting solution was filtered (5 lm) into an ultrafiltration cell and diafiltered (molecular weight cut off, 3000) with 10 exchange volumes of deionized water. The product, allyl-dextran, was then concentrated and lyophilized. The allyl-dextran was reacted with 7.5 g aminoalkyl thiol compound in 30 mL dimethylsulfoxide to produce an aminodextran conjugate. This reaction was initiated with 0.1 g ammonium persulfate and was performed under a nitrogen atmosphere. The volume of the reaction mixture after 3 h was doubled with deionized water, then, the solution was adjusted to pH 4 with sodium hydroxide (2.5N), and the product was diluted with 140 mL sodium acetate buffer (0.02 mol/L, pH 4). The product was then filtered (5 lm) into an ultrafiltration cell and dialyzed with five exchange volumes of deionized water. After concentration, the aminodextran conjugate was lyophilized. A sample was then assayed for the average number of amino groups per dextran, which was defined as the amine density. Aminodextran conjugate (0.5 g) and 35 mg $FeCl_3\_6H_2O$ were solved in 4 mL H_2O and nitrogen was flushed for 1.5 h. $FeCl_2\_4H_2O$ (14 mg) was added, followed by 100 lL aqueous ammonia in two portions while the mixture was kept under nitrogen. The solution turned black and was heated to 80 °C for 100 min. After the mixture was cooled to room temperature, the ammonia was removed by flushing the solution with nitrogen over 10 min. Freeze drying led to the desired particles (0.55 mg), which are stable at 4 °C for at least 1 year and were used for all further experiments. Titration of the resulting particle (18 mg) with 0.1 M HCl (0.85 mL, 85 mol) and bromophenol blue in acetone/H_2O (1:1, 10 mL) resulted in 3.3 mmol COO– g^{-1}. The particle size-distribution experiments were carried out as described above. 5-Aminosalicylic acid (0.33 mg, 0.23 mmol) and magnetic particles (1.0 mg, 3.3 mmol COO–, 20 equiv) were dissolved in 500 lL H_2O and the solution was shaken for 12 h at room temperature. To purify the product an ultrafiltration device was used for centrifugation and after concentration the sample was washed with H_2O (3 □ 2 mL). Size distribution experiments were carried out as described above. The morphology of the nanoparticles was determined usingscanning electron microscopy (SEM, Philips XL 30 scanning electron microscope, Philips, the Netherlands). The particles were coated with gold powder under vacuum before SEM. The crystallinity of the formed nanoparticles was followed with Philips X-ray diffractometer using Cu Ka radiation (k = 1.5406 Å) as s function of weight percent inorganic component. The Fourier transfer infrared (FT-IR) spectra of the nanoparticles were recorded on Perkin 810 spectrometer in KBr medium at room temperature in the region 4000–450 cm_1. X-ray diffraction patterns and particles size distribution of the aminodextran-coated iron oxide macro-molecule were measured by laser light scattering (Malvern Zetasizer ZS, Malvern UK). The samples were examined to determine the mean diameter and size distribution. The samples were prepared by suspending the freeze-dried nanoparticles in 10 mL deionized water (10 g/mL). Drug release from magnetic nanoparticles was carried out using a modified dissolution method. The media was a 0.05 M phosphate buffer solution. Nanoparticle powder (2 mg) was suspended in tubes containing buffer solution of pH 7.4 (10 mL) to simulate physiological pH. The tubes were placed in a shaker bath (Memmert WB14, Germany) at 37 °C.

RESULTS AND DISCUSSION

An important choice of dextran as the molecular backbone was based on practicality and availability. Therefore, active aminodextran molecules have been synthesized in 2-step process (Figure 1).

Figure 2 shows the SEM of aminodextran-coated iron oxide nanoparticles/ 5-aminosalicylic acid nanoparticles which were synthesized by chemical reaction. This nanoparticle is very sensitive to temperature. Scanning electron micrography images were obtained from a diluted solution of the nanocomposite particles. The white spots are drug nano particles. The SEM image shows the presence of 5- aminosalicylic acid spherical particles in polyfunctional macromolecule matrix, which are homogeneously distributed throughout the composites, which is also confirmed from 1H NMR studies. The ability of the aminodextran-coated iron oxide nanoparticles to form a complex with drugs depends on the nanoparticles and the electrostatic interactions between the nanoparticles and the drug. Therefore, it is possible to manipulate the incorporation process for a given drug by appropriate selection of the nanoparticlesand the surface functionality. One might expect that the 5-aminosalicylic acid with the carboxylic group may form a complex.

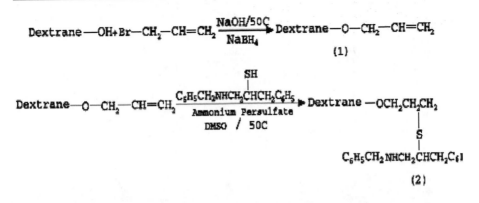

$$(1)$$

$$(2)$$

Figure 1. Covalent attachment of amino groups to dextran hydroxyl groups in 2-step process, which prevents dextran cross-linking. DMSO, dimethyl sulfoxide.

Figure 3 shows X-ray diffraction pattern of aminodextran-coated iron oxide nanoparticles. Diffraction of this macromolecule have a strong peak at about $2\Theta=26.46$, which is a characteristic peak of aminodextran-coated iron oxide nanoparticles. Studies on XRD patterns of nanoparticles are scarce in the literature.

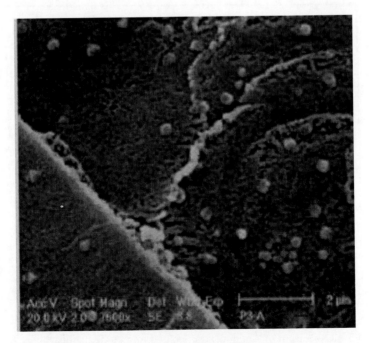

Figure 2. SEM of aminodextran- coated iron oxide nanoparticles.

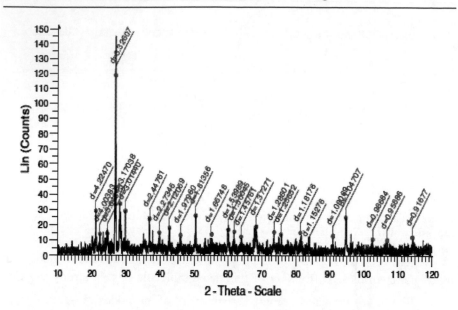

Figure 3. XRD pattern of aminodextran-coated iron oxide nanoparticles.

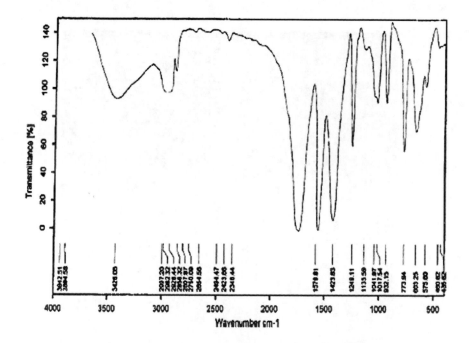

Figure 4. FT-IR spectrum of Aminodextran-coated iron oxide nanoparticles.

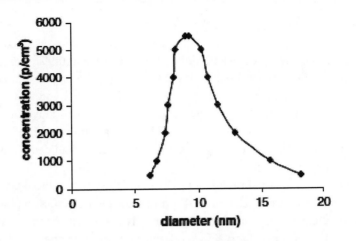

Figure 5. The particle size distribution of aminodextran-coated iron oxide nanoparticles.

Figure 4 showsthe FT-IR spectrum of aminodextran-coated iron oxide nanoparticles, where the % of transmittance is plotted as a function of wave number (cm^{-1}). The wide peak around 3411 cm^{-1} is attributing to the O–H stretching vibrations of aminodextran. The peaks at 1523 and 1714 cm^{-1} attribute to the COO- unsymmetrical and symmetrical stretching vibration, respectively. The mean diameter of each conjugate was measured by dynamic light scattering (UPA-150; Honeywell-Microtrac, Clearwater, FL).

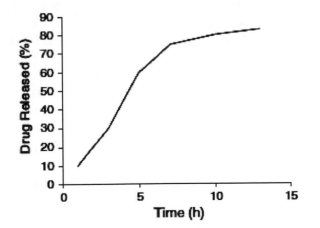

Figure 6. Drug release(%) to time of aminodextran-coated iron oxide nanoparticles.

Each conjugate was assayed for 10 min at a concentration >5 mg/mL of 0.9% saline. The refractor index of each sample was assayed (Fisher Scientific, Santa Clara, CA) and did not deviate from 0.9% saline. Latex particle standards of 3 different sizes gave weight-averaged diameters that were within 5% of their mean diameters (19 ± 1.5, 102 ± 3 and 993 ± 21 nm), which were calibrated by photon correlation spectroscopy or electron microscopy. The analyzer software did not assume a Gaussian size distribution. Mean molecular diameters with SDs were calculated from volume distribution data (Figure 5).

For the learning of nature and size effect of the drug in the drug delivery systems, the drug release behaviour of 5-amniosalicyclic acid as a pharmaceutically active compound has been studied. The concentration of the drug released at selected time intervals was determined by UV spectrophotometry, ($\lambda_{max} = 226$ nm), respectively. In order to study the potential application of nanoparticles containing 5-aminosalicylic acid as pharmaceutically active compounds, the drug release behavior of polymers have been studied under physiological conditions (Figure 6).

The concentration of drug in released at selected time intervals was determined by UV spectrophotometry. The decrease of particles size is an important parameter in the diffusion coefficient change. It appears that the degree of drug release polymer depends on their particle size. In the other hand, the chemical structure of the drug too is an important factor in the hydrolytic behavior of polymeric pro-drugs. The high different hydrolysis rate 5-aminosalicylic acid at pH 7.4 can be related to the functional groups along the drug. 5-Aminosalicylic acid contains both amine (basic) and carboxylic acid (acidic) functional groups. This factor ultimately results in an increase in hydrophilicity of 5-aminosalicylic acid at pH 7.4.

The average number of amino groups per dextran was measured in the following manner. The lyophilized dextran conjugate was dissolved in saline, and the amine concentration was measured by the trinitrobenzene sulfonate assay using hexylamine as a standard. The glucose concentration of the same sample was measured by the sulfuric acid method. The amino density was calculated by dividing the amine concentration by the glucose concentration and by multiplying the average number of glucose units per dextran.

CONCLUSIONS

Aminodextran-coated iron oxide nanoparticles are the first member of a new class of important agents based on macromolecular backbone with a high density of sites for the magnetic resonance imaging (MRI) reporters. This radiopharmaceutical is the first specifically designed anticancer drug carrier. Our long-term goal is to increase the pharmaceutical performance of the MRI technique. The result would be a wider dissemination, beyond academic centers and greater access for patients with cancer or melanoma.

REFERENCES

Chipowsky, S., & Lee, Y. C. (1973). Synthesis of 1-thioaldosides having an amino group at the aglycon terminal. *Carbohydrate Research,* 31, 339–346.

de Belder, D. (1996). Medical application of dextran and its derivatives. In S. Dumitru (Ed.), *Polysaccharides in medical applications* (pp. 505–524). New York, NY: Marcel Dekker.

Gershenwald, J. E., Tseng, C. H., & Thomson, W. (1998). Improved sentinel lymph node localization in patients with primary melanoma with the use of radiolabeled colloid. *Surgery,* 124, 203–210.

Giuliano, A. E., Kirgan, D. M., Guenther, J. M., & Morton, D. L. (1994). Lymphatic mapping and sentinel lymphadenectomy for breast cancer. *Annals of Surgery,* 220, 391–401.

Glass, L. F., Messina, J. L., & Cruse, W. (1996). The use of intraoperative radiolymphoscintigraphy for sentinel node diopsy in patients with malignant melanoma. *Dermatologic Surgery,* 22, 715–720.

Korosy, F., & Barczai-Martos, M. (1950). Preparation of acetobrome-sugars. *Nature,* 165, 169.

Krejcarek, G. E., & Tucker, K. L. (1977). Covalent attachment of chelating groups to macromolecules. *Biochemical and Biophysical Research Communications,* 77, 581–583.

Lee, Y. C., Stowell, C. P., & Krantz, M. J. (1976). 2-Imino-2-methoxyethyl-1-thioglycosides: New reagents for attaching sugars to proteins. *Biochemistry,* 15, 3956–3963.

Morton, D. L., Wen, D. R., & Wong, J. H. (1992). Technical details of intraoperative lymphatic mapping for early stage melanoma. *Archives of Surgery,* 127, 392–399.

Nugent, J., & Jain, R. K. (1984). Extravascular diffusion in normal and neoplastic tissues. *Cancer Research,* 44, 234–244.

Porter, C. J. H. (1997). Drug delivery to the lymphatic system. *Critical Reviews in Therapeutic Drug Carrier,* 14, 333–393.

Qu, T., Wang, Y., Zhu, Z., Rusckowski, M., & Hnatowich, D. J. (2001). Different chelators and different peptides together influence the in vitro and mouse in vivo properties of Tc. *Nuclear Medicine Communications,* 22, 203–215.

Shirakami, Y., Mtsumura, Y., Yamamichi, Y., Kurami, M., Ueda, N., & Hazue, M. (1987). Developing of Tc-DTPA-HAS as a new blood pool agent. *Japanese Journal of Nuclear Medicine,* 24, 475–478.

Thoren, L. (1978). Dextran as a plasma volume substitute. In G. A. Jamieson & T. J. Greenwalt (Eds.*), Blood substitutes and plasma expenders* (pp. 265–282). New York, NY: Alan R. Liss.

Tomalia, D. A., Baker, H., & Deward, J. (1985). A new class of polymers: Starburst dendritic macromolecules. *Polymer Journal,* 17, 117–132.

Vera, D. R., Buonocore, M. H., Wisner, F. R., Katzberg, R. W., & Stadalnik, R. C. (1995). A molecular receptor-binding contrast for magnetic resonance imagining of the liver. *Academic Radiology,* 2, 497–507.

Vera, D. R., Wisner, E. R., & Stadalnik, R. C. (1997). Sentinel node imaging via a nonparticulate receptor-binding radiotracer. *Journal of Nuclear Medicine,* 38, 530–535.

In: Modern Nanochemistry   ISBN: 978-1-61209-992-7
Eds: A. K. Haghi and G. E. Zaikov   © 2011 Nova Science Publishers, Inc.

Chapter 5

COLON DRUG DELIVERY SYSTEMS BASED ON CARBOXYMETHYL STARCH-CHITOSAN NANOPARTICLES

M. R. Saboktakin[1] and A. K. Haghi[2]

[1]Department of Nanotechnology, Baku State University, Azerbaijan
[2]University of Guilan, Iran

ABSTRACT

The purpose of this study was to examine chitosan(CS)-carboxymethyl starch(CMS) nanoparticles as drug delivery system to the colon. The 5-aminosalicylic acid (5-ASA) was chosen as model drug molecule.CS-CMS nanoparticles were formulated by a complex coacervation process under mild conditions.The influence of process variables, including the two ionic polymers, on particle size, and nanoparticles entrapment of 5-ASA was studied.In vitro release of 5-ASA were also evaluated, and the integrity of 5-ASA in the release fraction was assessed using sodium dodecyl sulfate-polyacrylamide gel electrophoresis. The release of 5-ASA from nanoparticle was based on the ion – exchange mechanism.The CS-CMS nanoparticles developed based on the modulation of ratio show promise as a system for controlled delivery of drug to the colon.

INTRODUCTION

Targeting of drugs specifically to the colon is advantageous in the treatment of diseases such as amoebiasis, Crohn's disease, ulcerative colitis, and colorectal cancer. In addition, it has shown great potential in the oral delivery of therapeutic peptides and proteins, which are unstable in the upper part of the gastrointestinal (GI) tract. The colonic region is recognized as having less diversity and intensity of enzymatic activities than stomach and small intestine [1]. Various strategies are available for targeting drug release selectively to the colon [2]. The designing of prodrugs is based on the concept of preventing the release of drugs in the stomach and small intestine and drug release is triggered by the use of specific property at the target site such as altered pH or high activity of certain enzymes in comparison to nontarget tissues [3]. Since it is known that azo function can be reduced in the colon [4], many novel polymers containing azo groups either in the polymeric backbone [5] or in the crosslinks [6] have been synthesized. To promote further selective degradation in the vicinity of the colonic environment, delivery systems have been designed that contain both pH-sensitive acidic monomers and degradable azo aromatic crosslinks [7]. Chitosan is a functional linear polymer derived from chitin, the most abundant natural polysaccharide on the earth after cellulose, and it is not digested in the upper GI tract by human digestive enzymes [8]. Chitosan is a copolymer consisting of 2-amino-2-deoxy-D-glucose and 2-acetamido-2-deoxy-D-glucose units links with β-(1-4)bonds. It should be susceptible to glycosidic hydrolysis by microbial enzymes in the colon because it possesses glycosidic linkages similar to those of other enzymatically depolymerized polysaccharides. The polysaccharide, on reaching the colon, undergoes assimilation by microorganisms or degradation by enzymes or break down of the polymer back bone leading to a subsequent reduction in molecular weight and thereby loss of mechanical strength and is unable to hold the drug any longer [9]. Chitosan has drawn attention for its potential to achieve site-specific delivery to the colon. Chitosan, a natural linear polyamine with a high ratio of glucosamine to acetyl-glucosamine units, is a weak base and carriers a positive charge. Its solubility is pH-dependent, and it reacts readily with negatively charged surfaces and materials, including polymers and DNA. Ionic gelation, complex coacervation, emulsion cross-linking, and spray-drying are methods commonly used for the preparation of chitosan nanoparticles [10]. Among those methods, ionic gelation and complex coacervation are mild processes occurring in a pure aqueous environment and are ideal for maintaining the in-process stability of drugs.

Ionic gelation and complex coacervation are very similar except that the former involves the gelation of chitosan using an electrolyte such as tripolyphosphate(TPP) [11], whereas the latter employs an oppositely chargedionic polymer such as alginate [12]. A new type of chitosan(CS) nanoparticles using carboxymethyl starch (CMS) as a polyanionic polymer was developed to achieve complex coacervation for the incorporation and controlled release of an antiangiogenesis hexapeptide [13, 14], this was the first report describing the use of CMS to formulate CS-based nanoparticles. Carboxymethyl starch (CMS) widely used in pharmaceuticals; however, it may need to be further modified for some special applications [15]. Among diverse approaches that are possible for modifying polysaccharides, grafting of synthetic polymer is a convenient method for adding new properties to a polysaccharide with minimum loss of its initial properties [16, 17]. We have synthesized CS-CMS nanoparticles as drug delivery system to the colon. The 5-ASA was chosen as model drug molecule [18-20].

Table 1. Compositions and Characteristics of Different CS-CMS Nanoparticles

Variables	Values	Degree of Swelling	Avargre Particle Size (mm)	Encapsulation Efficiency(%)
5-ASA	0	1.26±0.05	1.20±0.08	-
	5	1.14±0.20	1.68±0.08	72.5±2.34
	10	1.00±0.01	1.70±0.06	74.3±1.98
	15	0.90±0.09	1.92±0.04	82.1±2.54
	20	0.62±0.12	1.20±0.10	86.2±1.63
CS-CMS Nanoparticles (%wt/wt)	1	1.20±1.00	1.38±0.11	82.8±3.24
	2	1.30±1.21	1.75±0.09	83.4±1.25
	3	1.42±1.35	1.90±0.05	86.7±2.47
	4	1.60±1.40	1.87±0.02	87.9±2.35
Time(h)	10minute	0.60±0.52	1.90±0.12	72.4±1.47
	20 minute	0.73±0.14	1.87±0.21	87.1±1.98
	40minute	0.95±0.05	1.70±0.32	77.6±2.45
Drying(°C)	Lyophilization	0.96±0.50	1.32±0.12	68.2±4.35
	45	0.83±0.72	1.00±0.64	64.2±3.98

MATERIALS AND METHODS

Starch (M_w = 9500 g.mol^{-1}, 1g) and NaOH (1.2g) were suspended in isopropanol/ H_2O (85/15, 22ml) and heated to 60°C. Monochloroacetic acid (1.5g) was added slowly and the mixture was stirred for 2h at 60°C. After cooling to room temperature, the organic solvent was removed under reduced pressure and the aqueous phase was neutralized with acetic acid.Cold MeOH (30ml) was added and the solution was kept at 4°C overnight.After drying of the precipitate at high vacuum carboxymethyl starch(1)(1.5g)was obtained. Titration of starch-methylcarboxylate(1) (57mg) with 0.1MHCl (2.6ml, 0.26mmol) and bromophenol blue in acetone/H_2O (1:1, 10ml) resulted in 3.3mmol COO$^-$ g^{-1}. Therefore, on average, degree of substitution of carboxymethyl starch is 0.49 (DS=0.49). CS-CMS nanoparticles were prepared by the complex coacervation of CS and CMS. CS (0.25%) and CMS (1%)(wt/vol) solution were prepared by dissolving CS in aqueous acetic acid or CMS in water.The concentration of acetic acid was kept 1.75 times higher than that of CS in all cases to maintain the CS in the solution.The solution were then mixed with 5 mL of respective concentration of the CS solution under magnetic stirring (~ 200 rpm)at room temperature.The nanoparticle/ microparticle suspension was formed spontaneously.The mixture was stirred for a further 15 minutes.Both the pH and the particle size of the nanoparticle suspension were measured.

5-ASA was dissolved in methanol following which the CS-CMS nanoaprticles were added. The reaction mixture was stirred for 24 h in the dark, then evaporated using rotaevaporator to remove methanol. The traces were dried under vacuum in order to remove methanol completely. To these traces, deionized water was added. This solution was stirred in the dark for 24h. Then, the 5-ASA – CS-CMS nanoparticles were extracted. The solution was then filtered through PTFE membrane(Millix Millipore) of pore size 200nm, and then lypophilized to remove water.After approximately 180 min, the sample was sprayed into a liquid nitrogen bath cooled down to 77°K, resulting in frozen droplets. These frozen droplets were then put into the chamber of the freeze-dryer. Surface and shape characteristics of CS-CMS nanoaprticles were evaluated by means of a scanning electron microscope.The samples for SEM(Philips XL-30 E SEM) were prepared by lightly sprinkling the nanoparticles on a bouble adhesive tape, which stuck to an aluminum stub.The stubs were than coated with gold to a thickness of ~300Å using a sputter coater and viewed under the scanning electron microscope. Encapsulation efficiency (EE) is the amount of added drug (%) that is encapsulated in the formulation of the nanopartilces. The EE of drug from nanopartilces can be calculated in terms of the ratio of drug in the final formulation to the amount of added drug.An accurately weighed amount(100mg) of the formulation of nanopartilces were dispersed in 100 mL of Tris HCl buffer.The sample was ultrasonicated for 3 consecutive periods of 5 minutes each, with a resting period of 5 minutes each.It was left to centrifuged at 3000 rpm for 15 minutes. The concentration of 5-ASA in the decanted Tris HCl buffer and 2 washing solutions was determined by measuring the absorbance at 235 nm using a GBS Cintra 10-UV-Visible Spectrophotometer(Shimadzu, Japan).The determinations were made in triplicate, and results were averaged (Table 1).

Experimental data have been represented as the mean with standard deviation(SD) of different independent determinations.The significance of differences was evaluated by analysis of variance(ANOVA).Differences were considered statistically significant at $P < 0.005$.

RESULTS AND DISCUSSION

These nanoparticles had good spherical geometry.It is obvious that the surface of the nanoparticles shrank and a densely crosslinked gel structure was formed. This may explain the greater retardation of drug release from matrics of higher cross linker content.The average drug entrapment was found to be 81.21%±1.86% in the nanoparticles. Results of drug content and EE demonstrated that drug content increased from 10.23±0.50 mg/100mg to 26.24±0.43 mg/100mg of nanoparticles with increasing the amount of drug from 5% to 20% wt/wt. No signification increase in drug content was observed on further increasing the amount of drug, ie, above 15% wt/wt, which could be due to the limited solubility of the drug in DMF and that is endorsed from the presence of drug particles on the surface of the nanoparticles prepared with 20% of drug concentration.The percent EE was increased up to 86.70% ±2.47% with increasing polymer concentrations to 4%.Concentration of the cross-linking agent exhibited no significant effect on percent EE. SEM images of CS-CMS nanoparticles (Figure 1) show that nanoparticles have a solid and near consistent structure.

Figure 1. Morphology of 5-ASA loaded CS-CMS nanoparticles by emission scanning electron microscopy field.

Furthermore, the incorporation of 5-ASA into nanoparticles produced a smooth surface and compact structure. The particle size observed in SEM is smaller than that measured by the Zetasizer. This is because dried nanoparticles were used in SEM, whereas particles in the liquid dispersion was analyzed by the Zetasizer.CS-CMS nanoparticles are hydrophilic and would be expected to swell in water, thus producing a large hydrodynamic size when measured by the Zetasizer(Figure 2). 5-ASA – loaded CS-CMS nanoparticles were obtained spontaneously upon the mixing of the CMS aqueous solution (0.1%wt/vol) with the CS solution(0.1% wt/vol) under magnetic stirring, with 5-ASA dissolved in CS-CMS solution. The incorporation of 5-ASA into the CS-CMS nanoparticles resulted in a sharp increase in the particle size of the nanoparticle dispersion. The significant increases in particle size give a good induction of the incorporation of 5-ASA into CS-CMS Nanoparticles. A study was undertaken to investigate the effect of the order of 5-ASA mixing with CS and CMS. The data obtained show that the order of 5-ASA mixing had no effect on the size, entrapment efficiency, and yield of 5-ASA-loaded nano-particles.

The effect of drug concentration, CS-CMS concentration, GST concentration, and cross-linking time on in vitro drug release was also observed. In vitro drug release after 5 hours was 85.2%±4.0% in the case of nanoparticles having 15% drug, while it was 87.6%±3.1% for nanoparticles with 20% drug.The effect of CS-CMS on the release of drug was found to be meager.It is also observed that the amount of drug released from nanoparticles decreased on increasing crosslinking time. These properties are probably explained by the promotion of cross-links between CS-CMS chains and GST.

Freeze-drying of the samples resulted in larger and more porous nanoparticles compared with air-dried nanoparticles.Freeze-drying had the advantage of avoiding drug extraction by immediately freezing and removing the water present within the nanoparticles and a burst effect during the dissolution study(Figure 3). Conventional dissolution testing is less likely to accurately predict in vivo performance of colon delivery systems triggered by bacteria residing in the colon(because aspects of the colon's environment, ie, scarcity of fluid, reduced motility, and presence of microflora, cannot be simulated in conventional dissolution methods).

Hence, release studies were performed in an alternate release medium. The release profiles of 5-ASA-loaded CS-CMS nanoparticles were evaluated in water and a phosphate buffer, which was either in a different ionic strength or with saline, to study the underprinning mechanisms for drug release. The greatesr release for 5-ASA –loaded CS-CMS nanoparticles occurred in the release media of a high ionic strength.In contrast, a significantly small portion of 5-ASA was released in water over the release study period.The burst release was observed with nanoparticles, and it may have arisen from the desorption of loosely attached 5-ASA from the surface of the matrix polymers(Figure 4).

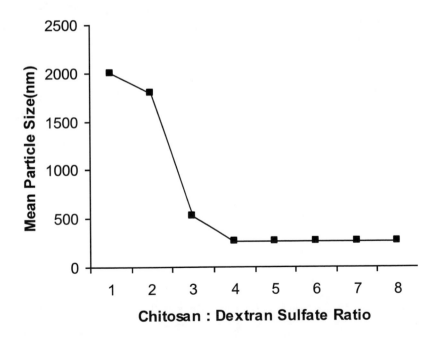

Figure 2. Influence of the CS: CMS ratio on the particle size.

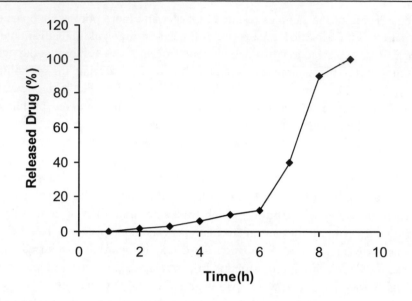

Figure 3. The in-vitro release profile of 5-ASA from CS-CMS nanoparticles in various simulated gastrointestinal fluids(n=3).

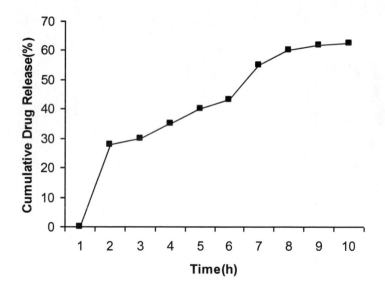

Figure 4. In vitro drug release from CS-CMS nanoparticles in SCF(pH 7.0)in presence 5-ASA.

CONCLUSIONS

CS-CMS nanoparticles containing 5-ASA were obtained by casting method with high loading efficiencies, and the chemical integrity of molecule was unaltered during preparation. Following an initial burst release, 5-ASA, being a hydrophobic drug, was not released further from the high molecular weight CS – CMS nanoparticles. Since films were observed to be susceptible to in vivo biodegradation mechanisms, 5-ASA release is expected in an appropriate environment that causes disruption of nanoparticles either in the form of erosion or degradation in vivo to facilitate release of the hydrophobic molecule.

REFERENCES

[1]   S.S. Davis; "Assessment of gastrointestinal transit and drug absorption", In: L.F. Prescott; W.S. Nimmo; eds.*Novel Drug Delivery and Its Therapeutic Application*, Chichester, UK:Wiley; (1990); 89-101.

[2]   M.K. Chourasia; S.K. Jain; "Pharmaceutical approaches to colon targeted drug delivery systems", *J Pharm Pharm Sci.*, 6; (2003); 33-66.

[3]   S. Davaran; J. Hanaee; A. Khosravi; "Release of 5-aminosalicylic acid from acrylic type polymeric prodrugs designed for colon-specific drug delivery", *J. of Control Release*, 58; (1999); 279-287.

[4]   E. Schacht; A. Gevaert; E.R. Kenawy; "Polymers for colon specific drug delivery ", *J. of Control Release*, 58; (1996); 327-338.

[5]   K.T. Chung; S.E. Stevens; C.E. Cerniglia; "The reduction of azo dyes by the intestinal microflora"; *Crit. Rev. Microbiol.*; 18; (1992); 175-190.

[6]   T. Yamaoka; Y. Makita; H. Sasatani; S.I. Kim; Y. Kimura; "Linear type azo-containing polyurethane as drug-coating material for colon-specific delivery : its properties degradation behavior and utilization for drug formulation"; *J. of Control Release*, 66; (2000); 187-197.

[7]   K.L. Shantha; P. Ravichandran; K.P. Rao; "Azo polymeric hydrogels for colon targeted drug delivery"; *Biomaterials.*; 16; (1995); 1313-1318.

[8]   G. Van den Mooter; C. Samyn; R. Kinget; "Azo polymers for colon-spicific drug delivery"; *Int. J. Pharm.*; 87; (1992); 37-46.

[9]   H. Ghandehari; P. Kopeckova; J. Kopecek; "In vitro degradation of pH sensitive hydrogels containing aromatic azo bonds"; *Biomaterials*; 18; (1997); 861-872.

[10] E.P. Kakoulides; J.D. Smart; J.Tsibouklis; "Azo crosslinked poly(acrylic acid) for colonic delivery and adhesion specificity synthesis and characterization"; *J. of Control Release*; 52; (2000); 291-300.

[11] I. Furda; "Aminopolysaccharides-their potential as dietary fiber.In: Furda I, ed.; *Unconventional Sources of Dietary Fiber, Physiological and In vitro Functional Properties*, Washington, DC:American Chemical Society; 14; (2000); 105-122.

[12] D.J. Ormrod; C.C. Holmes; T.E. Miller; "Dietary Chitosan inhibits hypercholesterolaemia and atherogenesis in the apolipoprotein E-deficient mouse model of atherosclerosis"; *Atherosclerosis*; 138; (2000); 329-334.

[13] A. Yamamoto; H. Tozaki; N. Okada; T. Fujita; "Colon specific delivery of peptide drugs and anti-inflammatory drugs using chitosan capsules"; *STP Pharma Science*; 10; (2000); 23-43.

[14] S.A. Agnihotri; N.N. Mallikarjuna; T.M. Aminabhavi; "Recent advances on chitosan-based micro- and nanoparticles in drug delivery"; *J. Control Release*; 100; (2004); 5-28.

[15] P. Calvo; C. Remunan-Lopez; J.L. Vila-Jato; M.J. Alonso; "Novel hydrophilic chitosan-polyethylene oxide nanoparticles as protein carriers"; *J. Appl. Polym. Sci.*; 63; (1997); 125-132.

[16] S. De; D. Robinson; "Polymer realationships during preparation of chitosan-alginate and poly-l-lisyne-alginate nanospheres"; *J. of Control Release*; 89; (2003); 101-112.

[17] Y. Chen; V. Mohanraj; J. Parkin; "Chitosan-dextransulfate nanoparticles for delivery of an anti-angiogenesis peptide"; *Int. J. Pept. Res. Ther.*; 10; (2003); 621-629.

[18] Z. Ma; H.H. Yeoh; L.Y. Lim; "Formulation pH modulates the interaction of insulin with chitosan nanoparticles"; *J. of Pharm Sci.*; 91; (2002); 1396-1404.

[19] Xu. Honghua; Li. Tiejing; "The analysis of boundary functions of CMS reaction factors"; *Nature and Science*; 3(2); (2005), 15-19.

[20] E. Jabbari; S. Nozari; "Swelling behavior of acrylic acid hydrogels prepared by γ- radiation crosslinking of polyacrylic acid in aqueous solution"; *Eur. Polym. J*; 36; (2000); 2685-2692.

In: Modern Nanochemistry
Eds: A. K. Haghi and G. E. Zaikov

ISBN: 978-1-61209-992-7
© 2011 Nova Science Publishers, Inc.

Chapter 6

LOCAL MUCOADHESIVE DRUG DELIVERY SYSTEMS BASED ON THIOLATED CHITOSAN-POLY(METHACRYLIC ACID) NANOPARTICLES

M. R. Saboktakin[1] and A. K. Haghi[2]

[1]Department of Nanotechnology, Baku State University, Azerbaijan
[2]University of Guilan, Iran

ABSTRACT

The main objective of this study was to develop a local, oral mucoadhesive metronidazole benzoate(MET) delivery system that can be applied and removed by the patient for the treatment of periodontal diseases. The results of present study revealed that the retention time of MET at its absorption site could be increased by formulating it into nanoparticles using thiolated chitosan(TCS) – poly(methacrylic acid)(PMAA). The nanoparticles of MET prepared from TCS-PMAA may represent a useful approach for targeting its release at its site of absorption, sustaining its release and improving its oral availability.

INTRODUCTION

TCS which are gaining popularity because of their high mucoadhesiveness and extended drug release properties. The derivatization of the primary amino groups of CS with coupling reagents bearing thiol functions leads to the formation of TCS [1]. The use of MET, which inhibits estrogen biosynthesis, is an attractive treatment for postmenopausal women with hormone – dependent breast cancer [2]. Since the early 1980s, the concept of mucoadhesion has gained considerable interest in pharmaceutical technology. If might open the door for novel, highly efficient dosage forms especially for oral drug delivery [3]. The most important goal of a cancer chemotherapy is to minimize the exposure of normal tissues to drugs while maintaining their therapeutic concentration in tomurs. Interestingly, nanoparticles (NPs) exhibits a significant tendency to accumulate in a number of tomurs after intravenous injection [4]. Hence, uptake and consequently bioavailability of the drug may be increased and frequency of dosing reduced with the result that patient compliance is improved [5, 6]. Various natural and synthetic polymers have been discovered as mucoadhesive excipients. Their mucoadhesive properties can be explained by their interaction with the glycoproteins of the mucus, based mainly on non-covalent bonds such as ionic interactions, hydrogen bonds and van der Waals forces [7, 8]. The biopolymer CS is obtained by alkaline deacetylation of chitin which one of the most abundant polysaccharides in nature [9]. Shell wastes of shrimp, lobster and crab are the main industrial sources of chitin [10]. The primary amino group accounts for the possibility of relatively easy chemical modification of CS and salt formation with acids. At acidic pH, the amino groups are protonated, which promotes solubility, whereas CS is insoluble at alkaline and neutral pH [11, 12]. Because of its favorable properties, such as enzymatic biodegradability, non-toxicity and biocompatibility CS has received considerable attention as a novel excipient in drug delivery systems, and has been included in the European Pharmacopoeia since 2002 [13]. The administration of nanoparticles will also provide the advantage of facilitating their injection through standard infiltration needles. So far, there was one published literature on MET nanoparticles prepared by direct precipitation technique [14]. Recently, it has been shown that polymers with thiol groups provide much higher adhesive properties than polymers generally considered to be mucoadhesive [15]. To increase patient complicance, to overcome the undesirable side effects, MET could be entrapped into biodegradable nanoparticles for sustained delivery so that it can inhibit estrogen biosynthesis for a prolonged time by virtue of increased local concentration of the drug at the receptor site [16, 17]. To date, three different TCS derivatives have been synthesized: CS-thioglycolic acid conjugates, CS-cysteine conjugates and chitosan-4- thio-butyl-amidine (CS-TBA) conjugates [18]. These TCS have numerous advantageous features in comparison to unmodified CS, such as significantly improved mucoadhesive and permeation enhancing properties [19]. The strong cohesive properties of TCS make them highly suitable excipients for controlled drug release dosage forms [20]. We have prepared MET-loaded TCS-PMAA nanoparticles by emulsion solvent evaporation technique to obtain smaller particle size with high entrapment efficiency and sustained release profile. Particle size, morphology, entrapment efficiency, drug – polymer interaction and *in vitro* release of MET-TCS-PMAA Nanoparticles were evaluated. The influence of % of drug (relation to polymer mass) on formulation performance including particle size, entrapment efficiency, *in vitro* release were investigated.

MATERIALS AND METHODS

The chemical modification of CS was performed as previously described. CS (500 mg) was dissolved in 50 mL of 1% acetic acid. In order to facilitate reaction with thioglycolic acid (TGA), 100 mg of ethyl-3-(3-dimethyla-minopropyl)carbodiimide hydrochloride (EDAC) was added to the CS solution. After EDAC was dissolved, 30 mL of TGA was added and the pH was adjusted to 5.0 with 3 N NaOH. The reaction mixture was stirred and left for 3 h at room temperature. To eliminate the unbonded TGA and to isolate the polymer conjugates, the reaction mixture was dialyzed against 5 mM HCl five times(molecular weight cut-off 10 kDa) over a period of 3 days in the dark, then two times against 5 mM HCl containing 1.0% NaCl to reduce ionic uninteractions between the cationic polymer and the anionic sulfohydryl compound. The thiolated chitosan with 1:1 molar ratios of MAA was polymerized at 60-70°C in a thermostatic water bath, bis-acrylamide as a cross-linking agent (CA), using persulfate as an initiator ([I] = 0.02 M) and water as the solvent (50 mL). The polymeric system was stirred by mechanical stirrer to sticky particles and it was separated from medium without solvent addition. All experiments were carried out in Pyrex glass ampoules. After the specific time (48 h), the precipitated network polymer was collected and dried in vacuum. TCS-PMAA (0.2 g) was dissolved in 15 mL 1% vol/vol acetic acid. For the loading of MET into polymeric matrix, 0.24 g MET was

suspended in the mixture. Before the addition of nanoparticles, a 200 μl
sample was taken and filtered using a low protein binding 0.22 μm PVDF
filter (Millipore, Bedford, MA) and then replaced with equal amounts of 1 ×
PBS, pH 7.4. An additional sample was taken in the same manner after
loading. The particles were collapsed using 10 mL 0.1 N HCl, filtered with
Whatman Grade 4 filter paper, and washed with 20 mL of deionized water.
After filtering, particles were frozen in a -80°C freezer and lyophilized at -
50°C under vacuum (LabConco Model 77500) for 24 hours. TCS-PMAA
nanoparticles were prepared as previously described and kept in a dry
environment until imaging. The nanoparticles were sprinkled onto an
aluminum stub that was covered with carbon tape. Excess nanoparticles were
removed by gently tapping the stub and the samples were sputtered coated
with a gold layer between 5 and 10 nm thick. Samples were imaged with a
SEM (Philips XL-30 E) at 10kV and a working distance of 7 mm.

Samples were allowed to swell on the surface of agar plates prepared in
simulated saliva (2.38 g Na_2HPO_4, 0.19g KH_2PO_4, and 8g NaCl per liter of
distilled water adjusted with phosphoric acid to pH 6.7) and kept in an
incubator maintained at 37°C ± 0.5°C. At preset time intervals (0.25, 0.5, 1, 2,
3, and 4 hours), samples were weighted (wet weight) and then left to dry for 7
days in a desicator over anhydrous calcium chloride at room temperature, then
the final constant weights were recorded.Water unptake (%) was calculated
using the following equation:

$$Water\ Uptake\ (\%) = \frac{w_w - w_f}{w_f} \times 100$$

Where w_w is the wet weight and w_f is the final weight. The swelling of
each sample was measured at least 3 times. Mucoadhesiveness was calculated
as the amount of mucin adsorbed by 2 mg of TCS-PMAA nanoparticles in a
certain time period. TCS-PMAA nanoparticle suspentions (4 mg/mL) were
mixed with type I-S mucin solution (0.5 and 1 mg/mL), vortezed, and
incubated at 37°C for 1, 6, 12, and 18 h. After adsorption, the suspensions
were centrifuged at 10, 000 × g for 10 min and free mucin was measured in the
supernatant by a colorimetric method using periodic acid/Schiff (PAS)
staining. Schiff reagent was prepared by diluting pararosaniline solution (40
g/L in 2 M HCl, Sigma) with water t to give a final concentration of 1.0%.
Sodium bisulfite (80 mg) was added to 5 mL of Schiff reagent and the

resultant solution was incubated at 37°C until it became colorless or pale yellow. Periodic acid solution was freshly prepared by adding 10 μL of 50% periodic acid to 7 mL of 7% acetic acid. Supernatants were mixed with 100μL of dilute periodic acid and incubated for 2 h at 37°C. Then, 100 μL of Schiff reagent was added at room temperature, and after 30 min the absorbance was measured at 560 nm. The amount of mucin adsorbed by the TCS-PMAA nanoparticles was determined by subtracting of concentration of mucin in solution after adsorption from that before. Mucin standards (0.1, 0.25 and 0.5 mg/mL) were measured by the same procedure and a standard calibration curve was prepared.

A definite weight range of 10-15mg of nanoparticles were cut and placed in a 1.5 mL capacity microcentrifuge tube containing 1mL of release medium of the following composition at 37°C: phosphate buffered saline (140mM, pH 7.4) with 0.1% sodium azide and 0.1% Tween 80.At predetermined time points, 100 μL of release medium was sampled with replacement to which 3 mL of scintillation cocktail was added and vortexed before liquid scintillation counting. The cumulative amount of MET released as a function of time was calculated.

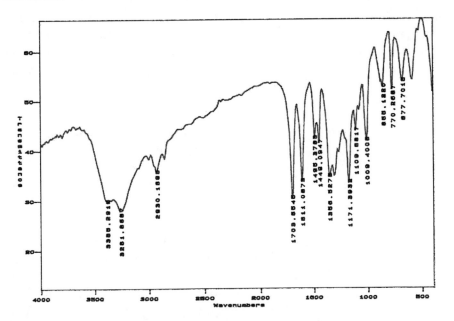

Figure 1. FT-IR spectra of *TCS-PMA* nanoparticles.

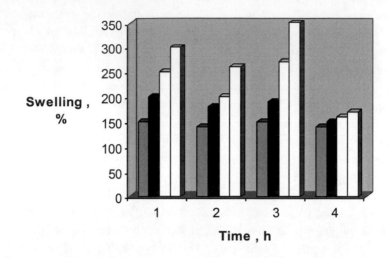

Figure 2. Percentage swelling measurement of *TCS-PMA* nanoparticles. Data are presented as mean ± SD(n=3).

Differential scanning calorimetry (DSC) studies were performed with a Mettler Toledo 821 thermal analyzer (Greifensee, Switzerland) calibrated with indium as standard. For thermogram acquisition, sample sizes of 1 to 5 mg were scanned with aheating rate of 5°C/min over a temperature range of 25°C to 300°C. In order to check the reversibility of transition, samples were heated to a point just above the corresponding transition temperature, cooled to room temperature, and reheated up to 300°C.

RESULTS AND DISCUSSION

The synthesized polymer was characterized by IR spectroscopy. The FT-IR spectrum of TCS-PMAA is shown in Figure 1. In the spectra of TCS −OH and −NH stretch were clearly seen at 3385.292 and 3261.866 cm^{-1}, respectively. Additional presence of amidine I and amidine II bonds are seen at 1703.855 and 1611.087 cm^{-1} corresponding to $\geq NH_2^+$ stretch and NH $\geq NH_2^+$, respectively, being two coupled vibrations. The presence of an additional bond at 1495.3783 and 1449.0847 cm^{-1} can be assigned for N-H bond of the salt $NH_2^+Cl^-$. Other characteristic peaks of CS O-H stretch, C-H stretch and C-O stretch were present at 3400-3600, 2930, and 1009-1171 cm^{-1}, respectively. This confirmed the synthesis of TCS. The spectrum of TCS was well correlated with reports by Matsuda *et al.* for TCS. Thiol content of TCS was found 214±52 µmol/g.

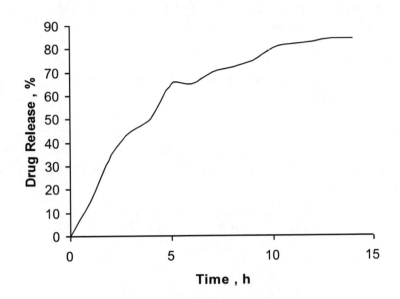

Figure 3. Percentage in vitro drug release of MET from *TCS-PMA* nanoparticles.

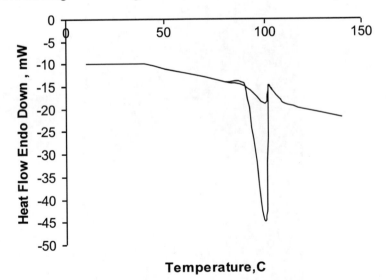

Figure 4. DSC thermograms of (a) *TCS-PMA* nanoparticles (b) MET Drug.

Figure 2 shows the percentage swelling of TCS-PMAA nanoparticles at different time intervals. The results revealed that nanoparticles swelled rapidly when immersed in phosphate buffer (pH 6.8). It is reported that adhesive properties and cohesiveness of mucoadhesive polymers are generally affected by their swelling behavior. Mucoadhesive nanoparticles are anticipated to take up water from the underlying mucosal tissue by absorbing, swelling and capillary effects, leading to considerable stronger adhesion. The percent swelling of nanoparticle was found to follow the rank order 248±18%, 198±15%, 279±26% and 164±15%, respectively. It was observed that TCS-PMAA nanoparticles swelled slowly and produced higher mucoadhesive strength. This is perhaps because slow swelling avoids the formation of over hydrated structure that loses its mucoadhesive properties before reaching the target.

Figure 5. SEM photomicrograph of *TCS-PMA* nanoparticles with *MET* drug.

Figure 3 shows the release of MET from mucoadhesive nanoparticles. Drug powder was completely released (95.3±4.1%) within 1 h. The time taken to release 75% of MET from TCS-PMAA nanoparticles was 5.0±0.6, 9.6±1.0, 4.2±0.2 and 5.3±0.5 h, respectively. The significantly higher time required by the TCS-PMAA nanoparticles to release MET was due to its better stability in acidic medium, which contributed significantly less amount of drug release during initial 1 h of dissolution (30.1±1.5% and 21.2±0.6% drug released from TCS-PMAA nanoparticles may be attributed to the higher solubility of CS in acidic medium. CS is soluble in acidic medium but crosslinking with glutaraldehyde through its amino group stabilized the nanoparticles matrix and provides the sustained release. The significantly lesser drug release from TCS-PMAA nanoparticles is due to the presence of disulphide bonds in nanoparticle matrix further stabilized the structure along glutaraldehyde as cross linking agent.

The thermal behavior of a polymer is important in relation to its properties for controlling the release rate in order to have a suitable drug dosage form. The glass transition temperature (Tg) was determined from the DSC thermograms (Figure 4). The higher Tg values probably related to the introduction of cross- links, which would decrease the flexibility of the chains and the ability of the chains to undergo segmental motion, which would increase the Tg values. On the other hand, the introduction of a strongly polar carboxylic acid group can increase the Tg value because of the formation of internal hydrogen bonds between the polymer chains.

The morphology of nanoparticles was examined by scanning electron microscopy(SEM, Philips XL-30 E). The nanoparticles were mounted on metal stubs using double – sided tape and coated with a 150 Å layer of gold under vacuum. Stubs were visualized under scanning electron microscope. Figure 5 shows the SEM of TCS-PMAA nanoparticles that synthesized by chemical reaction. This nanoparticles is very sensitive to the temperature that due to the interaction electron and sample. Scanning electron micrography images were obtained from a diluted solution of the MET particle. The white spots are MET nanoparticles. The SEM image shows the presence of MET spherical particles in polymer matrix, which are homogenenously distributed throughout the polymer, which is also confirmed from ^{1}H-NMR studies. As observed from SEM photomicrographs, the crystals of MET have a different appearance than recrystallized MET. These nanoparticles do not have clearly defined crystal morphological features in the SEM photomicrographs.

CONCLUSIONS

The results of present study revealed that the retention time of MET at its absorption site could be increased by formulating it into nanoparticles using TCS-PMAA. These nanoparticles prepared from TCS-PMAA showed the highest mucoadhesiveness. Further, they were observed to penetrate through the intestinal mucosa qualitatively better than the nanoparticles prepared from CS. These properties enabled sustained release of MET from nanoparticles prepared from TCS was maintained for 24 h. Hence, nanoparticles of MET prepared from TCS-PMAA may represent a useful approach for targeting its release at its site of absorption, sustaining its release and improving its oral availability.

REFERENCES

[1] Liu W.C.; Yao K.D.; (2002); "Chitosan and its derivaties-a promising non-viral vector for gene transfection"; *Journal of Control Release*; 83:1-11.

[2] Synder G.H.; Ready M.K.; Cennerazzo M.J.; Field D.; (1983); "Use of local electrostatic environments of cysteines to enhance formation of a desired species in a reversible disulfide exchange reaction"; *Journal of Biochemical Biophysical Acta*; 749: 219-226.

[3] Thanou M.; Nihot M.T.; Jansen M.; Verhoef J.C.; Junginger J.C.; (2001); "Mono-N-carboxymethyl chitosan (MCC), a polyampholytic chitosan derivative, enhances the intestinal absorption of low molecular weight heparin across intestinal epithelia in vitro and in vivo"; *Journal of Pharmaceutical Sciences*; 90:38-46.

[4] Baumann H.; Faust V.; (2001); "Concepts for improved regioselective placement of O-sulfo, N-acetyl, and N-carboxymethyl groups in chitosan derivatives "; *Journal of Carbohydrate Research;* 331: 43-57.

[5] Andreas B.S.; Krajicek M.E.; (1998); "Mucoadhesive polymers as platforms for peroral peptide delivery and absorption:synthesis and evaluation of different chitosan-EDTA conjugates "; *Journal of Control Release*; 50:215-223.

[6] Andreas B.S.; Hopf T.E.; (2001); "Synthesis and in vitro evaluation of chitosan-thioglycolic acid conjugates "; *Journal of Science pharmacy*; 69: 109-118.

[7]   Hornof M.D.; Kast C.E.; Andreas B.S.; (2003); "In vitro evaluation of the viscoelastic behavior of chitosan-thioglycolic acid conjugates "; *European Journal of Pharmaceutical Biopharmaceutics*; 55:185-190.

[8]   Andreas B.S.; Hornof M.; Zoidl T.; (2003); "Thiolated polymers-thiomers: modification of chitosan with 2-iminothiolane "; *International Journal of Pharmaceutics;* 260: 229-237.

[9]   Roldo M.; Hornof M.; Caliceti P.; Andreas B.S.; (2004); "Mucoadhesive thiolated chitosans as platforms for oral controlled drug delivery: synthesis and in vitro evaluation "; *European Journal of Pharmaceutical Biophamaceutics*; 57(1): 115-121.

[10]   Langoth N.; Guggi D.; Pinter Y.; Andreas B.S.; (2004); "Thiolated chitosan: in vitro evaluation of its permeation properties"; *Journal of Control Release*; 94(1): 177-186.

[11]   Kast C.E.; Valenta C.; Leopold M.; Andreas B.S.; (2002); "Design and in vitro evaluation of a novel bioadhesive vaginal drug delivery system for clotrimazole"; *Journal of Control Release*; 81:347-354.

[12]   Leitner V.M.; Marschutz M.K.; Andreas B.S.; (2003); "Mucoadhesive and cohesive properties of poly(acrylic acid)-cysteine conjugates with regard to their molecular mass"; *European Journal of Pharmaceutical Sciences*; 18:89-96.

[13]   Senel S.; Kremer M.; Kas S.; Wertz P.W.; Hincal A.A.; Squier C.A.; (2000); "Enhancing effect of chitosan on peptide drug delivery across buccal mucosa; *Journal of Biomaterials;* 21: 2067-2071.

[14]   Felt O.; Furrer P.; Mayer J.M.; Plazonnet B.; Buri P.; Gurny R.; (1999); "Topical use of chitosan in ophthalmology: tolerance assessment and evaluation of precorneal retention"; *International Journal of Pharmaceutics*; 180:185-193.

[15]   Felt O.; Buri P.; Gurny R.; (1998); "Chitosan: a unique polysaccharide for drug delivery"; *Journal of Drug Development and Industrial Pharmacy*; 24:979-993.

[16]   Kast C.E.; Frick W.; Losert U.; Andreas B.S.; (2003); "Chitosan – thioglycolic acid conjugate: a new scaffold material for tissue engineering "; *International Journal of Pharmaceutics;* 256: 183-186.

[17]   Shah R.B.; Ahsan F.; Khan M.A.; (2002); "Oral delivery of proteins: progess and prognostication "; *Critical Reviews in Therapeutics Drug Carrier System* "; 19: 135-169.

[18]   Coppi G.; Lannuccelli V.; Leo E.; Bernabei M.T.; Cameroni R.; (2001); "Chitisan-alginate microparticles as a protein carrier "; *Journal of Drug Development and Industrial Pharmacy*; 27: 393-400.

[19] Guggi D.; Kast C.E.; Andreas B.S.; (2003); "In vivo evaluation of an oral calcitonin delivery system for rats based on a thiolated chitosan matrix; *Journal of Pharmaceutical Research*; 20(12): 1989-1994.

[20] Bromberg L.E.; (2001); "Enhanced nasal retention of hydrophobically modified polyelectrolytes "; *Journal of Pharmacy and Pharmacology*; 53: 109-114.

In: Modern Nanochemistry
Eds: A. K. Haghi and G. E. Zaikov

ISBN: 978-1-61209-992-7
© 2011 Nova Science Publishers, Inc.

Chapter 7

SPECIFIC COLON DRUG DELIVERY SYSTEMS BASED ON CHITOSAN HYDROGELS CONTAINING 5-AMINOSALICYLIC ACID NANOPENDENTS

M. R. Saboktakin[1] and A. K. Haghi[2]

[1]Department of Nanotechnology, Baku State University, Azerbaijan
[2]University of Guilan, Iran

ABSTRACT

The main aim of this research was to develop and evaluate a multiparticulate system of Ac-poly(amidoamine)(PAMAM)(G4)–chitosan (CS) hydrogels exploiting pH-sensitive and specific bio-degradability properties for colon-targeted delivery of 5-aminosalicylic acid (5- ASA). All formulations were evaluated for particle size, encapsulation efficiency, swellability, and in vitro drug release. The size of the hydrogel was found to nanorange. The integrity of 5- ASA in the release fraction was assessed using sodium dodecyl sulfate-polyacrylamide gel electrophoresis. The CS–Ac-PAMAM hydrogel was developed based on the modulation of ratio show promise as a system for controlled delivery of drug.

INTRODUCTION

The biodegradable polymers are suitable biomaterials for the design of polymeric drug delivery devices for many classes of bioactive agents. These polymers have been used in various macromolecular architectures: linear, cross-linked and branched [1]. Many biodegradable polymers have been synthesized in recent times. One of the most important of dendrimer applications are the design of drug delivery systems [2]. Dendrimers offer a number of advantages compared to other architectural forms of polymers that have been used in drug delivery systems. They have narrow polydispersion, they are in the nanometer size ranges, which can allow easier passage across biological barriers, host-guest chemistry can take place either in the interior (binding groups in the interior of dendrimers are called endoreceptors) or on the periphery of the dendrimer (groups involved in completion chemistry on the periphery of the dendrimers are called exoreceptors) [3, 4]. Despite the extensive work on dendrimers, which has been grown potentially in recent years, their biopharmaceutical applications, much remains to be done to make their properties suitable for the intended applications [5].

Efforts are now directed at studying the biological and physicochemical properties of the dendrimers with a view to removing the limitations on their use. To change the bio-distribution patterns and to prolong circulation in the blood so as to facilitate targeting of specific tissues, hydrophilic polymers have been conjugated with dendrimers to shield positive charges and to create a steric barrier to reduce the potential for nonspecific interations such as opsonization [6, 7]. Kopecek has been developed a dendrimer-based system capable of achieving targeting with high specificity and low toxicity. In this research, we have been used from Ac-poly(amidoamine)(PAMAM) dendrimer (G4) as a base polymer for a drug delivery system.8 A side from other desirable properties of PAMAM dendrimer, their synthesis can be tailormade so as to influence the groups at the surface: full generations such as 1, 2 have amine functionalized surfaces, while half generations such as 1.5, 2.5 have carboxylic acid end groups at the surface. Efforts have been made to modify PAMAM dendrimer using chitosan (CS). However, either all the primary amines of the PAMAM dendrimer was shielded with chitosan (CS) or only 10% of the amino groups in PAMAM were covered [9, 10]. We have been hypothesized that it is possible to optimize the biological and physicochemical properties of PAMAM dendrimer modified with chitosan (CS) by graded changes in the chitosan (CS) block length in a manner analogous to the reports on CS [11]. Dendritic architecture and uniformly positioned functionality have been recently reported to carry the anti-inflammatory drug indomethacin for it is transdermal delivery as well as anticancer drugs [12]. Recent studies have been reported investigation on the effect of 4.0 generation (G) PAMAM dendrimer present in an anionic phospholipids composition, consisting of hydrogenated soyaphosphotidylcholine, cholesterol, diacetylphosphate and poly(ethylene glycol)-derivatized phosphotidylethanolamine, on the hydration and liquid crystalline structure formation [13]. In the present study, the 4.0 generation PAMAM dendrimer was exhaustively studied as controlled-release systems for parenteral administration of a model drug 5-aminosalicylic acid and analyzed using various release kinetic studies. This study gives us an insight about the biodistribution pattern of drug and it is localization at site of inflammation with PAMAM dendrimer [14, 15].

MATERIALS AND METHODS

The ratio between the acetic anhydride and the dendrimer was adjusted to achieve suitable degree of acetylation, with 100% primary amine groups converted. The amount of acetic anhydride was calculated based on the number of primary amines determined by potentiometric titration of ensure a 1:1 stoichiometric relationship between the acetic anhydride and the primary amino groups of G4 PAMAM dendrimer (When complete (100%) acetylation of dendrimer was planned, 20 mol% excess of acetic anhydride was used.) Trimethylamine (10% excess based on the amount of acetic anhydride) was added to quench acetic acid formed as a side product during the reaction. The reactions were carried out in a glass flask and anhydrous methanol solution at room temperature for 24 h. The reaction mixture was dialyzed first in phosphate buffer at pH 8.0 and then in deionized water. The purified samples were lyophilized and stored at -208°C. Ac-PAMAM/chitosan (CS) hydrogel was prepared using the method reported by Zhang et al., 16 with slight modifications. Chitosan (CS) and Ac-PAMAM were dissolved in different concentration (1–4%) of acetic acid (1%, w/v). 5-Aminosalicylic acid (5–20 mg) was dissolved separately in dimethyl formamide (DMF). Then, the chitosan (CS) solution (2.5 mL) and the drug solution (7.5 mL) were mixed together to obtain 10mL of chitosan (CS) drug solution. The chitosan (CS) drug solution was added dropwise (using a disposable syringe with a 22-gauge needle) into 40mL of sodium chloride-saturated Tris HCl buffer solution containing glutaraldehyde-saturated toluene (GST) in different concentration (1–3 mL). The samples were separated after 1 h of curing time and subsequently decanted, washed twice with 3mL of 0.05M Tris-HCl buffer, and samples were dried in vacuum oven at 408C. Then, 5-aminosalicylic acid (5-ASA) was dissolved in methanol following which the dendrimer/chitosan (CS) conjugate was added. The reaction mixture was stirred for 24 h in the dark, then evaporated using rotaevaporator to remove methanol. The traces were dried under vacuum in order to remove methanol completely. To these traces, deionized water was added. This solution was stirred in the dark for 24 h. Then, the drug–dendrimer complex was extracted, as dendrimer is soluble in water while 5-aminosalicylic acid (5-ASA) is not. The solution was then filtered through PTFE membrane (Millix Millipore, Amicon Bioseparation) of pore size 200 nm, and then lypophilized to remove water. After approximately 180 min, the sample was sprayed into a liquid nitrogen bath cooled down to 778K, resulting in frozen droplets. These frozen droplets were then put into the chamber of the freeze-dryer. In the freeze drying process, the products are dried by a sublimation of the water component in an iced solution. The drug-chitosan (CS)/dendrimer nanoparticles were obtained in the form of a brown powder. Surface and shape characteristics of chitosan–Ac- PAMAM hydrogels were evaluated by means of a scanning electron microscope (FEI-Qunta-200 SEM, FEI Company, Hillsboro, OR). The samples for SEM were prepared by lightly sprinkling the hydrogel on a bouble adhesive tape, which stuck to an aluminum stub. The stubs were than coated with gold to a thickness of -300A° using a sputter coater and viewed under the scanning electron microscope. Encapsulation efficiency (EE) is the amount of added drug (%) that is encapsulated in the formulation of the hydrogel. The EE of drug from hydrogel can be calculated in terms of the ratio of drug in the final formulation to the amount of added drug. An accurately weighted amount (100 mg) of the formulation of hydrogel was dispersed in 100mL of Tris HCl buffer. The sample was ultrasonicated for three consecutive periods of 5 min each, with a resting period of 5 min. It was left to centrifuged at 3000 rpm for 15 min. The concentration of 5-aminosalicylic acid (5-ASA) in the decanted Tris HCl buffer and two washing solutions was determined by measuring the absorbance at 235nm using a GBS Cintra 10-UV-Visible Spectrophotometer (Shimadzu, Japan). The determinations were made in triplicate and results were averaged. In vitro drug release studies were performed as per the scheme in different simulated fluids. Simulation of GI transit conditions was achieved by using different dissolution media. Simulated gastric fluid (SGF) pH 1.2 consisted of NaCl (0.2 g), HCl (7mL), and pesin (3.2 g); pH was adjusted to 1.2-0.5. Simulated intestinal fluid (SIF) pH 7.4 consist of KH_2PO4 (6.8 g), 0.2N NaOH (190 mL), and pancreatin (10.0 g); pH was adjusted to 7.4±0.1.

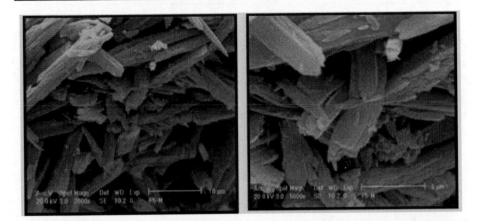

Figure 1. SEM of CS–PAMAM (G4)–(5-ASA) nanocomposite.

SIF pH 4.5 was prepared by mixing SGF pH 1.2 and SIF pH 7.4 in a ratio of 36:61. The experiment was performed with a continuous supply of carbon dioxide into dissolution media. The samples were withdrawn periodically and replaced with an equal amount of fresh dissolution media bubbled with carbon dioxide. The volume was made up to 10mL and centrifuged. The supernatant was filtered through Whatman filter paper (Dawsonville, GA), and drug content was determined spectrophotometrically at 235nm (UV 1601, Shimadzu, Japan).

RESULTS AND DISCUSSION

These hydrogels had good spherical geometry. It is obvious that the surface of the hydrogel shrank and a densely crosslinked gel structure was formed. This may explain the greater retardation of drug release from matrics of higher cross linker content. The average drug entrapment was found to be 83.46±1.56% in the hydrogel. Results of drug content and EE demonstrated that drug content increased from 13.23±0.50 mg/100mg to 28.44±0.43 mg/100mg of hydrogel with increasing the amount of drug from 5% to 20% w/w. No signification increase in drug content was observed on further increasing the amount of drug, that is, above 15% w/w, which could be due to the limited solubility of the drug in DMF and that is endorsed from the presence of drug particles on the surface of the hydrogel prepared with 20% of drug concentration. The percent EE was increased up to 89.32±0.20% with increasing polymer concentrations to 4%. Concentration of the cross-linking agent exhibited no significant effect on percent EE. FESEM images of CS/Ac-PAMAM hydrogel (Figure 1) show that hydrogel have a solid and near consistent structure. Furthermore, the incorporation of 5-ASA into the hydrogel produced a smooth surface and compact structure. The particle size observed in FESEM is smaller than that measured by the Zetasizer. This is because dried hydrogel was used in FESEM, whereas particles in the liquid dispersion was analyzed by the Zetasizer.

CS/Ac-PAMAM hydrogel are hydrophilic and would be expected to swell in water, thus producing a large hydrodynamic size when measured by the Zetasizer (Figure 2).

5-Aminosalicylic acid (5-ASA)-loaded hydrogel was obtained spontaneously upon the mixing of the Ac-PAMAM aqueous solution (0.1%, w/v) with the CS solution (0.1%, w/v) under magnetic stirring, with 5-ASA dissolved in CS–Ac-PAMAM solution. The incorporation of 5-ASA into the CS–Ac-PAMAM hydrogel resulted in a sharp increase in the particle size of the nanoparticle dispersion. The significant increases in particle size give a good induction of the incorporation of 5-ASA into CS–Ac-PAMAM hydrogel. A study was undertaken to investigate the effect of the order of 5-ASA mixing with CS and PAMAM. The obtained data show that the order of 5-ASA mixing had no effect on the size, entrapment efficiency, and yield of 5-ASA-loaded hydrogel. The effect of drug concentration, chitosan–Ac- PAMAM concentration, GST concentration, and cross-linking time on in vitro drug release was also observed. In vitro drug release after 5 h was 86.3_4.0% in the case of hydrogel having 15% drug, while it was 89.2±3.1% for hydrogel with 20% drug.

The effect of chitosan–Ac-PAMAM dendrimer on the release of drug was found to be meager. It is also observed that the amount of released drug from hydrogel decreased on increasing crosslinking time (Figure 2). The concentration of 5-aminosalicylic acid (5-ASA) in the decanted Tris HCl buffer and two washing solutions was determined by measuring the absorbance at 235nm using a GBS Cintra 10-UV-Visible Spectrophotometer (Shimadzu, Japan). The determinations were made in triplicate, and results were averaged (Figure 3).

These properties are probably explained by the promotion of cross-links between chitosan–Ac-PAMAM dendrimer and GST. Freeze-drying of the samples resulted in larger and more porous hydrogel compared with air-dried hydrogel. Freeze-drying had the advantage of avoiding drug extraction by immediately freezing and removing the water present within the hydrogel and a burst effect during the dissolution study (Figure 4).

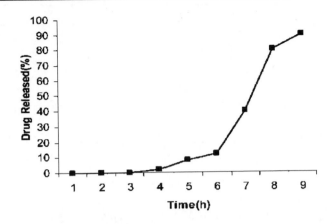

Figure 2. The *in vitro* release profile of 5-aminosalicylic acid from chitosan – PAMAM hydrogel in various simulated gastrointestinal fluids (n = 3).

Conventional dissolution testing is less likely to accurately predict in vivo performance of colon delivery systems triggered by bacteria residing in the colon (because aspects of the colon's environment, i.e., scarcity of fluid, reduced motility, and presence of microflora, cannot be simulated in conventional dissolution methods). Hence, release studies were performed in an alternate release medium. The resulting network polymers swell and become soft in solvents such asH2Oand most organic solvents without dissolving. To measure the swelling, preweighted dry drug-free hydrogels were immersed in various buffer solutions (pH 7.4 and pH 1) at 378C.

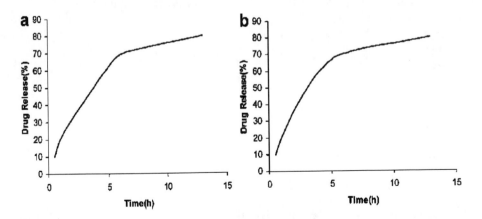

Figure 3. Release of drug (5-ASA) from (a) micro and 9b) nanopolymeric carriers as a function of time at 37°C.

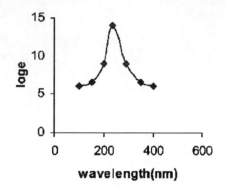

Figure 4. UV spectra of CS/Ac-PAMAM-5-ASA complex (λ_{max}=235 nm).

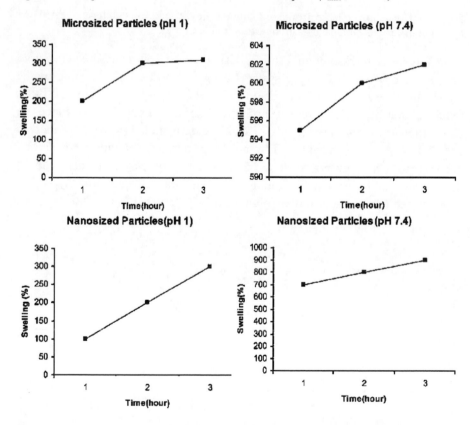

Figure 5. Time-dependent swelling behavior of micro and nanocarriers for 5-ASA drug model as a function of time at 37°C.

Then, excess water on the surface was removed with the filter paper, the weight of the swollen samples was measured at various time intervals. The procedure was repeated until there was no further weight increase (Figure 5).

The release profiles of 5-ASA-loaded hydrogel was evaluated in water and a phosphate buffer, which was either in a different ionic strength or with saline, to study the underprinning mechanisms for drug release. The greatest release for 5-ASA loaded hydrogel occurred in the release media of a high ionic strength. In contrast, a significantly small portion of 5-ASA was released in water over the release study period. The burst release was observed with hydrogel, and it may have arisen from the desorption of loosely attached 5-ASA from the surface of the matrix polymers.

CONCLUSIONS

Results of release studies indicate that CS/Ac- PAMAM hydrogel offer a high degree of protection from premature drug release in simulated upper GIT conditions. The present study reveals that the interaction of CS/Ac-PAMAM dendrimer conjugate with hydrophobic 5-aminosalicylic acid (5-ASA) molecules to bring it in its ionized state and hence enhance solubility. At the same time, dendrimers can localize the drug at the site of inflammation and the drug can provide effective pharmacological action. However, the potential role of our system in various other categories of the drugs for drug delivery is still under investigation. Also, the degree of substitution of Ac-PAMAM (G4) dendrimer-chitosan conjugates did not affect the encapsulation efficiency but affected the in vitro availability of the 5-aminosalicylic acid (5- ASA) at the early time of release.

REFERENCES

[1]   Kono K, Liu M, Frechet JMJ. 2000. Design of dentritic macromolecules containing folate or methotrexate residues. *Bioconjug Chem* 10:1115–1121.

[2]   Langer R. 1998. Drug delivery and targeting. *Nature* S392: 5–10.

[3]   Lestini BJ, Sagnella SM, Xu Z, Shive MS, Richter NJ. 2002. Surface modification of liposomes for selective cell targeting in cardiovascular drug delivery. *J Control Release* 78:235–247.

[4]   Malik N, Wiwattanapatapee R, Klopsch R. 2000. Dendrimers relation-
        ship between structure and biocompatibility in vitro, and preliminary
        studies on biodistribution of polyamidoaminedendrimers in vivo. *J
        Control Release* 65:133–148.

[5]   Milhem OM, Myles C, McKeeaown NB, Attwood D. 2000.
        Ployamidoamine starburst dendrimers as solubility enhancers. *Int J
        Pharm* 197:239–242.

[6]   Newkome GR, Moorefield CN, Vo¨glte F. 2001. *Dendrimers and
        Dendrons: Concepts, Syntheses, Applications.* Weinheim, Germany:
        Wiley-VCH.

[7]   Andrzej M, Thommey P, Baker J Jr. 2007. Dendrimers-Based Targeted
        Delivery of an Apoptotic Sensor in Cancer Cells. *Biomacromolecules*
        8:13–18.

[8]   Pretsch C, Seibl S. 1989. *Tables of spectral data for structure
        determination of organic compounds.* Berlin: Springer-Verlag.

[9]   Shultz L, Zimmerman S. 1999. Dendrimers potential drugs and drug
        delivery agents. *Pharm News* 6:25–29.

[10]   Socrates G. 1994. *Infrared characteristic group frequencies: Tables and
        charts.* New York: Wiley.

[11]   Tomalia DA, Baker H, Dewald J, Hall M, Kallos G, Martin S, Roeck J,
        Ryder J, Smith P. 1985. A new class of polymers: Starbust-dendritic
        macromolecules. *Polym J* 17:117–132.

[12]   Tomalia DA, Naylor AM, Goddasrd WA. 1990. Starburst dendrimers:
        Molecular-level control of size, shape, surface chemistry, topology and
        flexibility from atoms to macro-scopic matter. *Angew Chem Int Ed Engl*
        29:138–175.

[13]   Yoo H, Juliano RL. 2000. Enhanced delivery of antisense
        oligonuceotides with fluorophore-conjugated PAMAM dendrimers.
        Nucleic Acids Res 28:4225–4231.

[14]   Zhuo RX, Du B, Lu ZR. 1999. In vitro release of 5-flurouacil with
        cyclic core dendritic polymer. *J Control Release* 57:249–257.

[15]   Keunok YU, Paul SR. 1996. Light scattering and fluorescence
        photobleaching recovery study of polyamidoamine polymers in aqueous
        solution. *J Polym Sci B Polym Phys* 34:1467–1475.

[16]   Zhang H, Ibrahim A, Alsarra S, Neau H. 2002. An in vitro evaluation of
        a chitosan-containing multiparticulate system for macromolecule
        delivery to the colon. *Int J Pharm* 239: 197–205.

In: Modern Nanochemistry
Eds: A. K. Haghi and G. E. Zaikov

ISBN: 978-1-61209-992-7
© 2011 Nova Science Publishers, Inc.

Chapter 8

COLON DRUG DELIVERY SYSTEMS BASED ON DENDRITIC MACROMOLECULE NANOCARRIERS

M. R. Saboktakin[1] and A. K. Haghi[2]

[1]Department of Nanotechnology, Baku State University, Azerbaijan
[2]University of Guilan, Iran

ABSTRACT

Highly branched, functionalized polymers have potential to act as efficient drug carrier system. The aromatic polyether dendrimers are spherical, highly ordered, multibranched, nanometer-sized macromolecules having positively charged ether groups on the surface at physiological conditions. In this study, we have synthesized a kind of dendrimer / poly(2-hydroxy ethyl methacrylate) nanocomposite/5-aminosalicylic acid (5-ASA) for oral drug delivery. The aromatic polyether dendrimer (generation 2, hyperbranched polyether with -CH_2OH functionality, 3, 5-Dihydroxybenzoic acid core) was prepared from generation 2, hyperbranched polyether dendrimer with -$COOCH_3$ form in excellent yield. FT-IR, ^{1}H- NMR and DSC studied suggest that monomer predominantly forms an complex with polyether dendrimers because of the ionic interaction between the –CH_2OH end groups of dendrimer and the –OH group of 2-hydroxy ethyl methacrylate monomer.

INTRODUCTION

The most important characteristic of any drug is efficiency. This characteristic may often reduced because of the inability to deliver the drug to the specific cells or tissues [1, 2].After administration, the drug may pass through different physiologic barriers and /or pathways, decreasing the actual amount drug that reaches the site.In the search for an ideal carrier system, the dendrimers may have significant potential. Dendrimers are synthetic macromolecules with a well-defined globular structures [3]. The need for advanced materials with improved and new properties for a variety of technological applications has created a demand for both new forms of matter and for polymers that have highly controlled molecular architectures [4, 5].The established approach to dendritic macromolecules has traditionally involved a divergent process in which growth is started from a polyfunctional core and continued outwards in a stepwise manner that affords larger and larger macromolecules as the process is continued [6, 7]. The fundamental attribute of the convergent approach is that it begins at what will be the periphery of the molecule, proceeding inwards [8, 9]. It is this feature more than any other that allows for unparalleled control over molecular architecture [10, 11]. We have investigated the potential of dendrimers amd hyperbranched polymers as drug carrier system using poly(2-hydroxy ethyl methacrylate), since the methodologies for evaluation of the cellular activity of dendrimer are well known [12, 13].In this paper, we have explored the interaction of the poly(2-hydroxy ethyl methacrylate), with aromatic polyetheric dendrimer. The nature of the interaction was characterized by FT-IR and ^{1}H-NMR spectroscopy [14, 15].

MATERIALS AND METHODS

A mixture of the G2-COOCH$_3$ dendrimer(AB8, heptamer) (0.8 g, 0285 mmol) in dry THF (100 ml) and dropwise to a suspension of LiAlH$_4$ (2.5 g, 60 ml) in dry THF (50ml). After reflux for 1h, the solution was treated with aqueous NaOH (1M, 15ml), filtered and evaporated. The residue was chromatographed on silica gel with dichloromethane as the eluent to give the heptameric alcohol as a colorless glass(0.768 g, 96%).R$_f$ = 0.91. A mixture of the G2-CH$_2$OH dendrimer(AB8, heptamer) (0.17 g, 0.279 mmol), 2-hydroxy ethyl methacrylate (0.176 g, 0.127 mmol), diethyl azo dicarboxylate(0.05 g,

0.317 mmol) triphenyl phosphine (0.083 g, 0.317 mmol) was dissolved in dry THF. The reaction mixture was stirred for 24 h in the dark, then evaporated using rotaevaporator to remove methanol. The traces were dried under vacuum in order to remove methanol completely. To these traces, deionized water was added. The residue was chromatographed on silica gel with dichloromethane-petroleum ether (3:1) as the eluent to give the pure conjugate. The 2-hydroxy ethyl methacrylate –dendrimer conjugate obtained was in the form of a yellow powder(R_f = 0.92, 68.1%). A mixture of the 2-hydroxy ethyl methacrylate - G2-CH$_2$OH dendrimer(AB8, heptamer) conjugate (4.83 g, 0.37 mmol), azo bis isobutyronitryle (AIBN) (61m g, 0.37 mmol), DMF (10 ml) at 6162 °C for 24 h was stirred for 24 h. The mixture was then filtered through PTFE membrane(Millix Millipore) of pore size 200nm, and then lyophilized to remove water. The 2-hydroxy ethyl methacrylate –dendrimer nanocomposite was obtained in the form of a white powder(m.p. 253-254 °C, 73.0%). 0.5 g of polymer bonded drugs (PBDs) containing 5-ASA was dispersed with stirring in 25 ml deionised water. After approximately 180 min, the PBDs were sprayed into a liquid nitrogen bath cooled down to 77° K, resulting in frozen droplets. These frozen droplets were then put into the chamber of the freeze-dryer. In the freeze-drying process, the products are dried by a sublimation of the water component in an iced solution. Figures 4 show scanning electron microscope (SEM) of nano polymer bonded drugs.

RESULTS AND DISCUSSION

Figure 1 shows the FT-IR spectrum of the G2-COOCH$_3$ dendrimer(AB8, heptamer) where the % of transmittance is plotted as a function of wave number (cm^{-1}). The characteristic FT-IR peaks at 3078, 3053, 2961, 1718 cm^{-1} are due to the presence of =CH bond stretching vibrations, the aliphatic CH bond in protected groups (t-butyldimethylsilyloxy) and carbonyl (C=O) group, respectively.

Figure 2 shows the FT-IR spectrum of the G2-CH$_2$OH dendrimer(AB8, heptamer). The characteristic FT-IR peaks at 3408, 2960, 2932, 2860, 1544 cm^{-1} are due to the presence of OH phenolic group, the aliphatic and aromatic =CH bond in protected methyl groups and C=C bond of aromatic group, respectively.

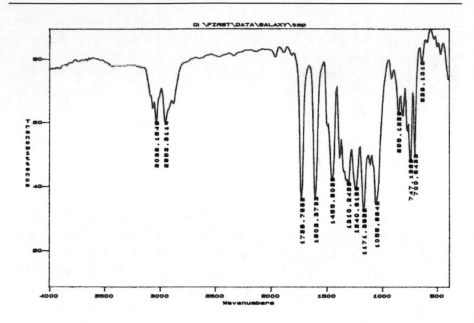

Figure 1. FT-IR spectrum of [G2] –COOCH₃ Dendrimer.

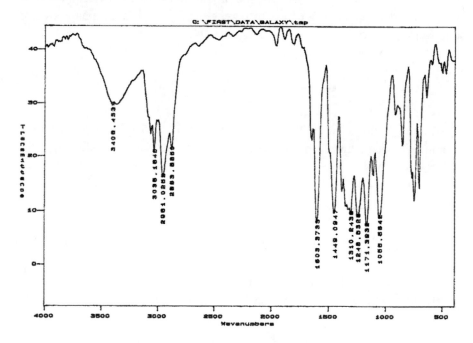

Figure 2. FT-IR spectrum of [G2] – CH₂OH Dendrimer.

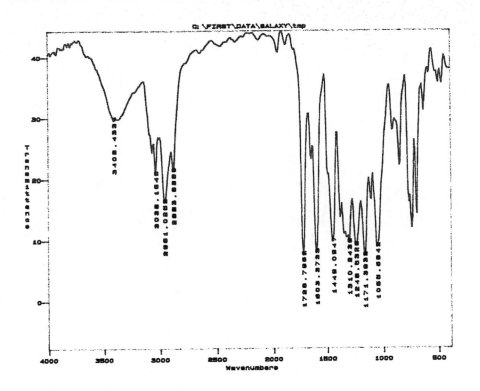

Figure 3. FT-IR spectrum of [G2] – CH₂OH Dendrimer/poly(2-hydroxy ethyl methacrylate)/5-ASA nanocomposite.

Figure 4. SEM of [G2] – CH₂OH Dendrimer/2-hydroxy ethyl methacrylate/5-ASA nanocomposite.

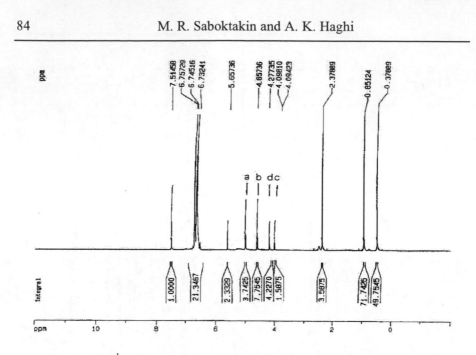

Figure 5. 400MHz ^{1}H-NMR spectrum of [G2] – CH$_2$OH Dendrimer/2-hydroxy ethyl methacrylate conjugate(CDCl$_3$).

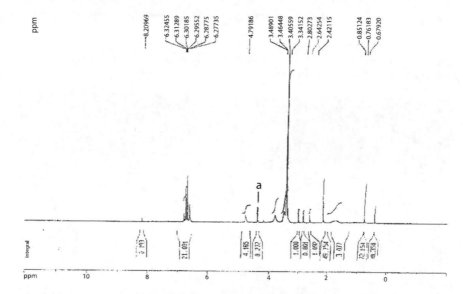

Figure 6. 400MHz ^{1}H-NMR spectrum of [G2] – CH$_2$OH Dendrimer/poly(2-hydroxy ethyl methacrylate) nanocomposite (DMSO-d6).

Also, Figure 3 shows the FT-IR spectrum of the G2-CH$_2$OH dendrimer (AB8, heptamer)/ poly(2-hydroxy ethyl methacrylate) nanocomposite.The characteristic FT-IR peaks at 3408, 2960, 2932, 2860, 1544, 1726cm^{-1} are due to the presence of OH phenolic group, the aliphatic and aromatic =CH bond in protected methyl groups and C=C bond of aromatic group and carbonyl (C=O) group of 2-hydroxy ethyl methacrylate, respectively.

Figure 4 shows the SEM of aromatic polyether dendrimer / poly(2-hydroxy ethyl methacrylate) nanocomposite with 5-aminosalicylic acid (5-ASA) as drug model. This nanocomposite is very sensitive to the temperature that due to the interaction electron and sample. Scanning electron micrography images were obtained from a diluted solution of the nanocomposite particles. The white particles are 5-ASA nanoparticles. The SEM image shows the presence of 5-aminosalicylic acid (5-ASA) particles in polyfunctional dendrimeric matrix, which were distributed throughout the nanocomposite, which is also confirmed from ^{1}H-NMR studies [15]. The ability of the dendrimer to form a complex with 2-hydroxy ethyl methacrylate monomer depends on the core- surface groups of dendrimer, electrostatic interactions between the dendrimer and the monomer, and the ability of the drug to form a conjugate with the dendrimer through chemical bonding. Therefore, it is possible to manipulate the incorporation process for a given monomer by appropriate selection of the dendrimer and the surface functionality.One might expect that the 2-hydroxy ethyl methacrylate with the carboxylic group may form a complex with surface OH groups of polyether dendrimer.

Figure 5 shows the 400 MHz ^{1}H-NMR spectrum of of [G2] – CH$_2$OH Dendrimer/2-hydroxy ethyl methacrylate conjugate in which three regions can be seen. The silane group 48 and 72 protons are in the regions of 0.37 and 0.85 ppm, respectively. The resonance of CH$_2$ protons (c group), CH$_2$, 4 protons (d group), CH$_2$ 8 protons (Heptamer protons), benzylic protons of heptamer(a group), vinylic CH$_2$ protons are 4.09, 4.15, 4.27, 4.86, 5.66 ppm, respectively. The resonances for the aromatic protons of the monomer units at dendrimer occur in the region 6.73 ppm separate resonances are observed in the appropriate ratio for each layer of monomer units. The solvent peak occur in the region 7.51 ppm.

Figure 6 shows the 400MHz 1H-NMR spectrum of of [G2] – CH2OH Dendrimer/poly(2-hydroxy ethyl methacrylate)nanocomposite in which three regions can be seen. The silane group 48 and 72 protons are in the regions of 0.10 and 0.85 ppm, respectively. The resonance of methyl group, polymer CH$_2$ protons, CH$_2$O protons, CH$_2$ 8 protons (Heptamer protons), CH$_2$ protons of heptamer(a group), are 1.68, 2.27, 4.17, 4.65, 4.79 ppm, respectively. The resonances for the aromatic protons of the monomer units at dendrimer occur in the region 6.28-6.32 ppm separate resonances are observed in the appropriate ratio for each layer of monomer units. The DMSO peak occur in the region 3.34-3.48 ppm.

The thermal behavior of a polymer is important in relation to its properties for controlling the properties of nanocomposite. The glass transition temperature (Tg) was determined from the DSC thermograms. The values are given in Table1. The higher Tg values probably related to the nanocomposite size, which would decrease the flexibility of the chains and the ability of the chains to undergo segmental motion, which would increase the Tg values. On the other hand, the introduction of a strongly phenolic OH group can increase the Tg value because of the formation of internal hydrogen bonds between the polymer chains.

Nano and micro polymer bonded drugs (50 mg) were poured into 3 mL of aqueous buffer solution (SGF: pH 1 or SIF: pH 7.4). The mixture was introduced into a cellophane membrane dialysis bag. The bag was closed and transferred to a flask containing 20 mL of the same solution maintained at 37° C. The external solution was continuously stirred, and 3 mL samples were removed at selected intervals. The removed volume was replaced with SGF or SIF. Triplicate samples were used. The sample of hydrolyzate was analyzed and the quantity of 5-ASA was determined using a standard calibration curve obtained under the same conditions(Figure 7).

Results of drug content and encapsulation efficiency(EE) demonstrated that drug content increased from 9.24±0.81 mg/100mg to 25.70±0.43mg/100mg with increasing the amount of drug from 5% to 20% wt/wt. No increase in drug content was observed on further increasing the amount of drug, above 15% wt/wt, which could be due to the limited solubility of drug in solvent and that is endorsed from the presence of drug particles on the surface of the nanocomposite with 20% of drug concentration.In the case of nanocomposite, nearly 80% to 90% of the drug was released in the initial 8 to 10 hours.This situation is not acceptable for drugs that are required to be released locally in the colon.The effect of drug concentration, nanocomposite concentration and cross-linking time on in vitro drug release was observed.In vitro drug release after 10 hours was 84.2%±4.1% in the microsized particles and pH =7.4 having 15% drug, while it was 89.2%±3.78% for nanosized particles with 20% drug.The effect of nanocomposite concentration on the release of drug was found to be meager. It is also observed that the amount of drug released from nanocomposite decreased on increasing cross-linking time.These properties are probably explained by the promotion of cross-links between nanocomposite chains.Freeze-drying of the samples resulted in larger and more porous nanocomposite and slightly faster disintegration times compared with air-dried nanocomposite. Freeze-drying had the advantage of avoiding drug extraction by immediately freezing and removing the water present within the nanocomposite by sublimation.When air-dried, the nanocomposite was exposed to water within for longer periods of time.In the second of this investigation cross-linked nanocomposite was as nanosized paricles. The relatively higher drug release after 10 hours reveals a signification increasing. This increase in drug release suggests that most of the drug has dissolved and leached out through in nanocomposite.The drug released during 10-hours release rate studies, is because of the presence of nanosized paricles.

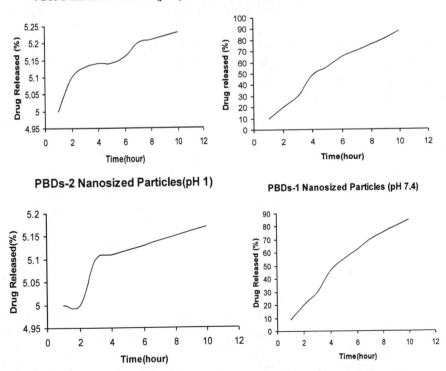

Figure 7. Release of 5-ASA drug from micro and nano polymeric carriers as a function of time at 37°C.

CONCLUSIONS

The ability of the dendrimer to form a nanocomposite with 2-hydroxy ethyl methacrylate monomer as a nano drug system was explored using 5-aminosalicylic acid (5-ASA) as drug model.The nature of 2-hydroxy ethyl methacrylate -dendrimer interaction were explored using FT-IR, ^{1}H-NMR and SEM. Our studies suggest that the aromatic polyether dendrimer may predominantly form a conjugate with the -OH group of 2-hydroxy ethyl methacrylate. This nanocomposite is stable in deionized water and methanol and can be use as drug carrier system. Current studies are exploring the complexation/conjugation ability of these dendrimers to a wide variety of drugs carrier.

REFERENCES

[1]   Hawker, C.J.; Frechet, J.M.J.; "The convergent synthesis of dendritic polyethers based on 3, 5- dihydroxybenzyl alcohol ", *J.Chem. Soc. Chem. Commun.*, 1990, 112, 7638-7644.

[2]   Wooly, K.L.; Hawker, C.J.; Frechet, J.M.J.; " Alternative convergent synthesis of dendritic polyethers "; *J.Am.Chem.Soc.*; 1991, 113, 4252-4258.

[3]   Leon, W.J.; Kawa, M.; Frechet, J.M.J.; "The new synthesis of dendritic polyethers ", *J.Am. Chem. Soc.*; 1996, 118, 8847-8852.

[4]   Craig, J.; Hawker, C.J.; Frechet, J.M.J.; "Preparation of polymers with controlled molecular architecture.A new convergent approach to dendritic macromolecules "; *J.Am.Chem.Soc.*; 1990, 112, 7638-7647.

[5]   Craig, J.; Hawker, C.J.; Frechet, J.M.J.; " Monodispersed dendritic polyesters with removable chain ends: a versatile approach to globular macromolecules with chemically reversible polarities "; *J.Chem.Soc.Perkin Trans.* 1992, 1, 2459-2469.

[6]   Forier, B.; dehaen, W.; " Alternative convergent and accelerated double-stage convergent approaches towards functionalized dendritic polyethers "; *Tetrahedron,* 1999, 55, 9829-9846.

[7]   Saboktakin, M.R.; Maharramov, A.; Ramazanov, M.A.; "Synthesis and characterization of aromatic polyether dendrimer / Mesalamine (5-ASA) nanocomposite as drug carrier system "; *J. Am.Sci.*; 2007, 3(4), 47-51.

[8]   Saboktakin, M.R.; Maharramov, A.; Ramazanov, M.A.; "Poly (amidoamine) (PAMAM) /CMS Dendritic nanocomposite for controlled drug delivery"; *J. Am.Sci.*; 2008, 4(1), 48-52.

[9]   Nicolaos, A., P.; Kelly, B.K.; Madeline, T.L.; Lowman, A.M.; "Poly(ethylene glycol)-containing hydrogels in drug delivery"; *J. of Controlled Release*; 1999, 62, 81-87.

[10] Padias, A.B.; Hall, H.K., Tomalia, J.; McConnell, D.A.; "Synthesis of dendritic macromolecules via a convergent approach"; *J. Org.Chem.*; 1987, 52, 5305-5311.

[11] Naylor, A.M.; Goddard, W.A.; Kiefer, G.E.; Tomalia, D.A.; "A new convergent approach to dendritic macromolecules"; *J.Am. Chem. Soc.*, 1989, 111, 2339-2344.

[12] Aharoni, S.M.; Crosby, C. R.; Walsh, E. K..; "Size – exclusion chromatography of dendritic macromolecules"; *Macromolecules*, 1982, 15, 1093-1102.

[13] Sonke, S.; Tomalia, D.A.; "Dendrimers in biomedical applications— reflections on the field "; *Advan. Drug. Delivery. Rev.*; 2055, 57, 2106-2129.

[14] Tomalia, D.A.; "The convergence of quantized dendritic building blocks/architectures for applications in nanotechnology"; *Chemistry Today*; 2005, 23(6), 41-45.

[15] Petkova, V.; Parvanova, V.Tomaila, D.L.; "3D Structure of dendritic and hyper –branched macromolecules by X-ray diffraction"; *Soild State comm.. J.*; 2005, 134, 671-675.

In: Modern Nanochemistry
Eds: A. K. Haghi and G. E. Zaikov

ISBN: 978-1-61209-992-7
© 2011 Nova Science Publishers, Inc.

Chapter 9

RHEOLOGICAL PROPERTIES OF POLY(METHYL METHACRYLATE) NANOCOMPOSITES FOR DENTURE SCIENCES

M. R. Saboktakin[1] and A. K. Haghi[2]

[1]Department of Nanotechnology, Baku State University, Azerbaijan
[2]University of Guilan, Iran

ABSTRACT

Hybrid materials, which consist of organic – inorganic materials, are of profound interest owing to their unexpected synergistically derived properties. Aluminium oxide nanoparticle/polymer composites have been produced using a one – system polymer synthesis. The linear polymer, poly(methyl methacrylate)(PMMA, M_W=15, 000g/mol) is applied for the stabilization of aluminum oxide(Al_2O_3) nanoparticles. The fourier transfer infrared(FT-IR) analysis data and scanning electron microscopy(SEM) image revel that the core shell structure of aluminum oxide/PMMA nanocomposite have been synthesized. The ratio of concentration of the capping polymer material to the concentration of the aluminium oxide precursor could control the size of aluminium oxide nanoparticles. With specific concentration of the reductant, the core – shell nanostructure could be fluctuated in order.

INTRODUCTION

Polymers are of profound interest to society and replacing metals in diverse fields of life, which can be further modified according to modern application. The Organic –inorganic hybrid materials are hi-tech because they can present simultaneously both the properties of an inorganic molecule besides the usal properties of polymer [1]. Poly(methyl methacrylate)(PMMA) has excellent physical properties and clearly defined polymerization process that is easy for modification.Many attempts have been made to modify PMMA taking advantage of the broad scope of modification available in polymer chemistry [2, 3]. In previous study, the experimental resin had a negative charge incorporate by copolymerization of methacrylic acid to methyl methacrylate [4]. Results showed that the adhesion of C. albicans decreased significantly as the amount of MMA increased in vitro [5]. A signification decrease in adhesion to the resin samples existed when the Al_2O_3 was present at 10% of the modified PMMA. The synthesis of metallic nanoparticles has become an extremely interesting topic in the field of material science, due to the wide range of optical and electric properties that are accessible in nanometer-sized regime [6, 7].

The unique properties of such nanopatricles are attributed to quantum confinement or surface effects, which become operative when the particle is too small to exhibit bulk behavior [8]. Under certain conditions, nanoparticles of Al_2O_3 can be induced to undergo self – organization into three-dimensional superlattices [9]. Such nanoparticles assemblies open the door to "tunable" materials, in which optical and electronic properties are dependent on both initial cluster sizes and the manner in which clusters organize to form larger structures [10]. For example; surfactants act as control agents in mineralization processes through the formation of complexes with inorganic ions [11. In addition, during nanoparticle synthesis, surfactants have been used as surface capping agents in order to stabilize the nanoparticles and to prevent coagulation [12]. In the present paper, we have reported the in-situ synthesis of quantum-confined Al_2O_3 nanoparticles and poly(methyl methacrylate) copolymer hybride material.PMMA copolymer was employed due to its great thermal transfiguration property [13]. The glass transition temperature (T_g) of PMMA is about 85-110 °C, which is dependent on molecular weight and film thickness. The T_g of copolymer is relatively lower than those of other types of copolymers such as polycarbonate (PC), acrylonitrile-butadiene-styrene(ABS) and so forth. This advantage makes it easier to change the conformation of PMMA/Al_2O_3 nanocomposites with relatively lower temperature [14].This synthesis method is in contrast to previous examples of micelles-forming systems, in which aggregates were prepared from solutions containing multi components of stabilizers [15].

MATERIALS AND METHODS

In a 50 ml flask, 35 g of the monomer mixture was stirred with 1.2 g of benzoyl peroxide was poured into a 250 ml flask containing 1% poly(vinyl alcohol) at pH 3 and stirred well to prevent separation of two layers, and the temperature was recorded. The reaction was allowed to continue for 15 minutes after the rise in temperature ceased. The polymer beads were filtered, washed with distilled water, and dried. The same procedure was followed except after half an hour of reaction the sol was added. The sol was prepared by mixing Al_2O_3 nanoparticles solution(2 ml), deionized water(0.4ml) and HCl(0.04ml, used to catalyze the hydrolysis)in a beaker and pre-hydrolyzed in air and then the mixture was sonicated for 15 minute to facilitate the conversion of materials. The pH value of 1.3 was used to obtain a stable solution.Furthermore, this homogeneous mixture was added drowise over stirring to avoid local in homogeneities. The particle size –distribution experiments were carried out as described above.This experiment was repeated for four groups of modified PMMA. Four plates per each experimental group were fabricated.Polymerization of the resin was carried out in water at $55\pm1°C$ in a pressurized chamber(22 psi) for 15 minutes.Each plate was divied into five equal strips producing 25 samples per experimental group.These oversized strips were milled to the digitally calibrated dimensions [10mm(w) × 65mm(L) × 2.5mm(D)] and polished to minimize surface roughness. The samples were washed with distilled water to remove any residual monomer and then stored in distilled water at 37°C for 50 ± 2 hours before testing.

RESULTS AND DISCUSSION

Figure 1 illustrates the micrographs of the obtained composites revealing that their production was successfully achieved yielding materials with particles well dispersed within the matrices. Results show the micrograph of virgin polymers and it can be seen that the distribution of size is not uniform and the particles size varies. They range from 3-12 μm in size and their chain formation is clearly visible from the micrograph. The virgin polymer also exhibits porous nature while the pores disappear in the composite structure. This result illustrates that the nanoparticles are intercalated into the structure of polymer. In this Figure, SEM picture shows Al_2O_3 nanoparticles on the surface of the copolymer surface. The hydrophilic Al_2O_3 particles on the surface of the copolymer, hydrophilic due to the hydroxyl groups on the Al_2O_3 combined with inherent surface roughness impart hydrophilic nature, according to Cassie's equation. During the reaction, the hydrophilic Al_2O_3 particles migrated to the polymer water interface due to Van dr Waal's attraction.

The micrograph shows a distribution of two groups of about 1-2 μm and 0.5 μm Al_2O_3 particles, which are spherical in shape. Another evidence for the particle coating was provided by FT-IR analysis. After washing with ethanol to remove most of the free PMMA polymer, a small amount of associated PMMA might remain. Figure 2 shows a FT-IR spectrum of a nanocomposite that was prepared from Al_2O_3 and PMMA isolated. The vibration band of carbonyl($\upsilon_{C=O}$) at $1780 cm^{-1}$ is characteristic of the PMMA branches. It is clear from the comparision of this spectrum with the free PMMA spectrum that PMMA is present in appreciable quantity of the composite material. The interaction between PMMA and the Al_2O_3 surface is probably due to a hydrophobic interaction.

The PMMA polymer exhibits hydrophobic characteristics. Polymers allowed good interactions both with the Al_2O_3 surface and the Al_2O_3 precursor for obtaining stable colloids.In the present work, in the absence of PMMA/PMAA copolymer, precipitation occurs immediately or shortly. Utilizing a 3-point flexural test, the samples were mounted in a calibrated Instron Universal Testing Machine (Instron Corp., Canton, MA). Each plastic strip was supported on each end by metal rollers 50mm apart.A centrally located rod applied a load until fracture occurred at a uniform crosshead speed of 2.5 mm/min. Force-deflection curves and a complete stress versus strain history for each test were obtained. An instron computer program was used to calculate the tranverse strength, transverse deflection, flexural strength, and modulus of elasticity from the data curves along with the means and standard deviations for each experimental group. A representation of the difference in mean transverse strength is shown in Figure 3.

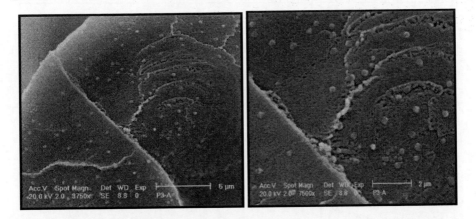

Figure 1. SEM of PMMA/Al₂O₃ nanocomposites.

This nanocomposite shoed the highest mean force required to fracture the specimens.As the amount of MMA increased, the transverse strength decreased.The transverse deflection measurements and the mean values are shown in Figure 4.

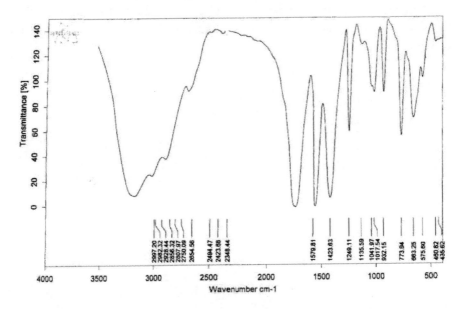

Figure 2. FT-IR spectra of PMMA/Al₂O₃nanocomposites.

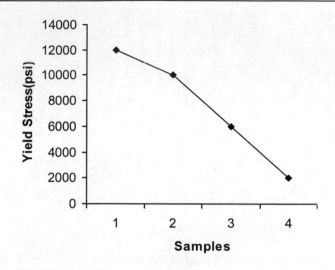

Figure 3. Transverse strength or force at fracture for 4 sample groups.

Results show that as the amount of MMA increased, the transverse deflection decreased, indicating a decrease in its flexibility. The Figure 5 shows the mean and standard deviation values for flexural strength for each of the experimental groups. As the the amount of MMA increased, the flexural strength decreased.

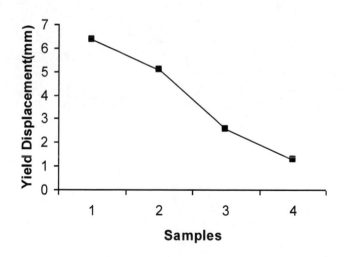

Figure 4. The transverse deflection for 4 sample groups.

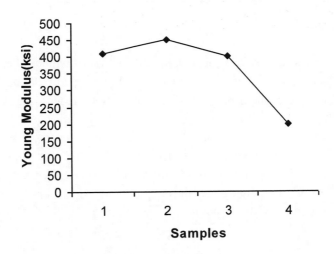

Figure 5. The flexural strength for 4 sample groups.

Figure 6 illustrates the diffractograms of PMMA and PMMA-Al$_2$O$_3$ nanocomposites in the 2θ range between 5 and 90 degree, which are similar and without any sharp diffraction peaks confirming their non-crystalline nature.The interlayer spacing of system was determined by the diffraction peak in the X-ray method, using the Bragg equation:

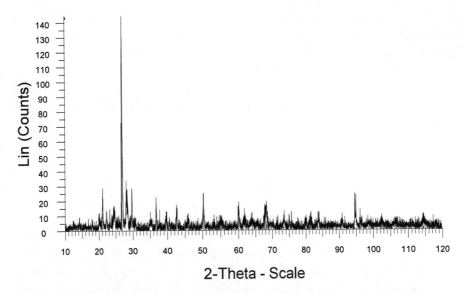

Figure 6. XRD pattern of a) PMMA(red) b)PMMA-Al$_2$O$_3$ nanocomposite(black).

$$\lambda = 2d\sin\theta$$

Where d is the spacing between diffractional lattice planes, θ the diffraction position while λ the wavelength of the X-ray(1.5405 Å).PMMA is known to be an amorphous polymer and shows three broad peaks at 2θ values of 21°, 27° and 29° (d spacing around 4 Å, 2.94 Å and 2.79 Å), with their intensity decreasing systematically.The shape of the first most intense peak reflects the ordered packing of polymer chains while the second peak denotes the ordering inside the main chains. The addition of Al_2O_3 nanoparticles do not induce any crystallinity in these polymers. This also explains the homogeneous nature of these samples.

CONCLUSIONS

We have studied that a nanometer PMMA copolymer network could be formed by Al_2O_3 nanoparticles as a template system.The present work is significant for several resons: (1) synthesis of Al_2O_3 nanoparticles within the self assembly of block PMMA copolymers in organic solvent, (2) assemblies of nanoparticles within a polymer matrix, with spatial confinement at the nanometer scale, (3) employment of the in-situ synthesis strategy for the synthesis of organic-inorganic hybrid nano network structures.Controling the surface properties of nanomaterials is a major technological research area encompassing studies in the pharmaceutical, mining, semiconductor, biological and medical fields. This work demonstrates a method to generate network structures and represents a powerful and general strategy for highly functional materials.

REFERENCES

[1]   Vallittu, P. K.; (1996); "Dimensional accuracy and stability of polymethyl methacrylate reinforced with metal wire or with continuous glass fiber," *The Journal of Prosthetic Dentistry*, 75(6), pp. 617–621.

[2]   John, J.; Gangadhar, S. A. and Shah, I.; (2001); "Flexural strength of heat-polymerized polymethyl methacrylate denture resin reinforced with glass, aramid, or nylon fibers, " *The Journal of Prosthetic Dentistry*, 86(4), pp. 424–427.

[3]   Park, S.E.; periathamby, A.R.; Loza, J.C.; (2003); "Effect of surface-charged poly(methyl methacrylate) on the adhesion of Candida albicans", *Journal of Prosthodontics*, 12(4), pp.249-254.

[4]   Waltimo, T.; Tanner, J.'Vallittu, P.'Haapasalo, M.; (1999); "Adherence of Candida albicans to the surface of poly(methyl methacrylate-E glass fiber composite used in dentures", *International Journal of Prosthodontics*, 12(1), pp.83-86.

[5]   Langer, L.; Stieneker, F.; Lambrecht, G., kreuter, E.; (1997); "Methyl methacrylate sulfopropylmethacrylate copolymer nanoparticles for drug delivery, *International Journal of Pharmaceutical*, 158(1), 211-217.

[6]   Barbeau, J.; Seguin, J.; Goulet, P.; (1993); "Reassessing the presence of candida albicans in denture-related stomatites"; *Oral Radiology Endodontics*; 2003'95, 1; 51-59.

[7]   Regezi, J.A.; Sciubba, J.; "*White lesions on oral patology*, 93-135.

[8]   Budtz Jorgensen, E.; (1974); "The significance of Candida albicans in denture stomatitis"; *Scandinavian Journal of Dental research*, 82; 2, 151-190.

[9]   Torres, S.R.; Peixoto, C.B.; Caldas, D.M.; (2003); "Clinical aspects of Candida species carriage in saliva of xerotomic subjects"; *Medical Mycology*, 41, 5, 411-415.

[10]   Slots, J.; Taubman, M.A.; (1992); "Infections of oral Microbilogy and Immunology "; 476-499.

[11]   Darwazeh, A.M.G.; Refai, S.; Mojaiwel; (2001); "isolation of Candida species from the oral cavity and fingertrips of complete denture wearers"; *The journal of Prosthetic Dentistry*, 86, 4; 420-423.

[12]   Jagger, R.G.; Hugget, R.; (1990); "The eefect of cross-linking on sorption properties of a denture –base materials "; *Dental Materials*, 6, 4, 276-278.

[13]   O'Brien, W.J.; (2002); "Polymers and polymerization: denture base polymers"; *In Dental Materials and Their Selection*, 74-89.

[14]   Ladizesky, N.H.; Ho, C.F.; Chow, T.W.; (1992); "Reinforcement of complete denture bases with continous high performace polyethylene fibers"; *The Journal of Prosthetic Density*, 68, 6, 934-939.

[15]   Samaranayake, L.P.; MacFarlane; (1980); "An in-vitro study of the adherence of Candida albicans to acrylic surfaces; *Archives of oral Biology*; 25, 8-9, 603-609.

In: Modern Nanochemistry   ISBN: 978-1-61209-992-7
Eds: A. K. Haghi and G. E. Zaikov   © 2011 Nova Science Publishers, Inc.

Chapter 10

THE CHARACTERIZATION OF ELECTROMAGNETIC SHIELDS BASED ON HYBRIDE POLYANILINE NANOCOMPOSITES

M. R. Saboktakin[1] and A. K. Haghi[2]
[1]Department of Nanotechnology, Baku State University, Azerbaijan
[2]University of Guilan, Iran

ABSTRACT

Conducting polymers have been widely used because of their lower density as well their good environmental stability as in the case of polyaniline (PAN). The doping process of PAN, which turn it conductive, is an important stage for the electromagnetic shields performance. For this, sulfonic acid is normally used because of the thermal behavior offered to the final material. Camphor sulfonic acid has been used because of the metallic characteristic gave to the PAN, increasing the conductive propertiy of the polymer. In this research, we have been studied the effects of camphor sulfonic acid as doping material at the different temperatures. Also, we have synthesized a conductive foam as electromagnetic shield.

INTRODUCTION

The electromagnetic radiation interference is one of the by-product of the rapid proliferation of electronic devices. These are undesiered conducted or radiated electrical disturbance including transients which can interfere with the operation of electrical or electronic components. The nano-structured materials have attraction for microwave radiation absorbing and shielding materials in the GHz frequency range due to their unique chemical and physical properties [1]. The volume to weight ratio of shielding material is very important in microwave absorbing materials for lightweight and strong absorption properties. The use of plastic materials to the housing of computer and electronic devices has been grown very rapidly due to their advantages over metals, like light weight, design flexibility, low cost and easy to mass production. As such plastic casing of electronic equipment do not provide protection from external field [2]. Composites with discontinous conducting fillers, such as metal particles are extensively employed in electromagnetic interference (EMI) shielding. Nowdays the conducting polymers offers a great technological application potential in several areas [3], can be cited: static films for transparent packaging of electronic components, electromagnetic shielding, rechargeable batteries, light – emitting diodes, nonlinear optical devices, sensor for medicine and pharmaceutics apparatus, membranes for separation of gas mixture, protection against corrosion, conducting paints and glues and others. The most important application of these polymers is like microwave adsorbing materials [4]. These polymers are generally prepared by adding fillers in a polymeric matrix. One very common way, among the several methods for preparing conducting polymer blends or composites, is by mechanical mixing of the components [5]. Great interest has been focused on PAN and polyurethane resin (PU) matrix, within the field of conducting polymers, due to important characteristics that it presents: its conductive form has excellent chemical stability combined with relatively easy, inexpensive and with high – yield. These blends may combine the desired properties of two components, the electrical conductivity of PAN/PU resin with the physical and mechanical properties of the polymeric matrix [6-7]. We have been studied some microwave absorption properties of doped PAN/PU resin with SiO_2 [8]. Frequency conductivity and dielectric relaxation measurements especially have been proven to be valuable in giving additional information on the conduction mechanism that conductivity measurement alone does not proposed [9, 10]. In this work a new potential use SiO_2 has been proposed.

The high dielectric constant SiO_2 has been used for the synthesis of PAN/PU resin/ SiO_2 nanocomposites. Here we have reported our observation of the morphology and conducting properties of doped PAN/PU resin / SiO_2 nanocomposites, synthesized by chemical oxidation process [11, 12].

MATERIALS AND METHODS

Aniline (AR grade) was purified by distillation before use and ammonium persulfate [$(NH_4)_2S_2O_8$], HCl were used as received. Add 0.558 gram of aniline to HCl 1M solution(1000 cm^3) to form PANI and 15% of SiO_2 nanoparticles was added to this solution with vigorous stirring in order to keep the SiO_2 nanoparticles suspended in the solution. Dissolve 0.006 mol ammonium persulfate [$(NH_4)_2S_2O_8$] with 10 ml distilled water. Mix these two solutions to start the polymerization reaction. The polymeric matrix is a dispersion of particle of 100-200 nm in diameter. The reaction mixture was agitated continuously for 8 h. In this process, the solution was always kept at 5 °C. The precipitate formed was collected by filtration and wash with distilled water and acetone until the filtrate became colourless. In the mean time, the precipitate was also tested to ensure that there were no free sulfate ions in the filtrate by using $BaCl_2$ solution. Any unreacted aniline in the solution formed was removed by acetone. After washing, the precipitate, freeze drying led to the desired particles (0.55mg), which are stable at 4°C for at least one year and were used for all further experiments. The powder PAN – HCl was neutralized with NH_4OH 1M obtaining PAN-emeraldine base. PAN emeraldine base doping with camphor sulfonic acid was carried out reactive processing doping. In this method, process into an internal mixer chamber of a hake rheometer 600. The reactive processing with camphor sulfonic acid was made at 100 °C.The ratio of PAN emeraldine base: camphor sulfonic acid at 50 rpm for 10 min. Blends of PU bi component with 15% w/w of PU – comphor sulfonic acid were prepared using a mechnical mixer at 1500 rpm for 50-60 min. A shield was obtained applying the mixture on an aluminium flat plate to get homogeneous film with thickness of 2 mm.

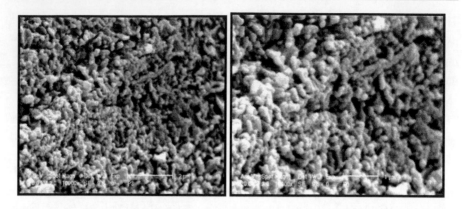

Figure 1. SEM of doped polyaniline with SiO_2 nanoparticles foams.

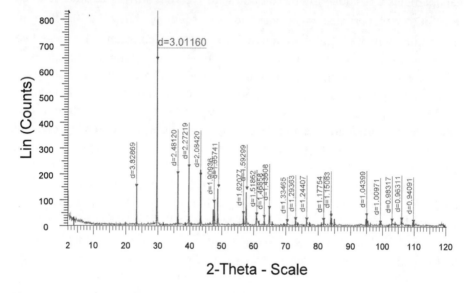

2-Theta - Scale

Figure 2. XRD pattern for conductive polyaniline foam with SiO_2 (15%) nanoparticles.

RESULTS AND DISCUSSION

SEM of doped PAN with SiO_2 nanoparticles foams synthesized by chemical oxidative is shown in Figure 1. Doped PAN with SiO_2 nanoparticles foams is very sensitive to the temperature, Due to the interaction electron and sample. Scanning electron micrography images were obtains from a diluted solution of the nanocomposite particle. The white spots are SiO_2 nanoparticles.

The SEM image shows the presence of spherical SiO_2 particles in doped PAN foams, which are homogenenously distributed throughout the composites, which is also confirmed from XRD studies. A very high magnification of SEM image shows the presence of spherical SiO_2 particles in PAN, which are homogeneously distributed throughout the composites, which is also confirmed from XRD studies. It is for the first time such a beautiful distribution of cenospheres is observed which looks as if the beads are floating over the water surface. These ceospheres show a large variation in their dimensions. Since the particles of SiO_2 are spherical in shape, the observed porosity in these composites is less than the other PAN composites.

The crystallinity of the formed composites was followed with X-Ray diffraction (XRD) as s function of weight percent inorganic component. Studies on XRD patterns of PAN are scarce in the literature. Figure 2 shows the XRD pattern for PAN: SiO_2 (15%). The diffraction pattern of PAN: SiO_2 nanocomposite shows a peak at about $2\Theta = 29.89°$.

Figure 3a shows the FT- IR spectrum of doped PAN, where the % of transmittance is plotted as a function of wave number (cm^{-1}). The characteristic FT-IR peak at 1579 and 1423 cm^{-1} are due to the presence of quinoid and benzenoid rings, respectively and are clear indication of these two states in the polymer chain. Also, The peaks at 1135, 1249, $1579 cm^{-1}$ are due to the C-N bond stretching vibration, N=O, respectively. Also, Figure 3b shows the FT-IR spectrum of doped PAN in the presence of SiO_2 particles exhibit new absorption peaks distinctly at 1579, 1423, 1249, 1135 and 773 cm^{-1} which are assignable to the presence of various metal oxides in the composite. The FT-IR spectra of other nanocomposites (PAN with 15% SiO_2) do not show much variation in the characteristic peaks.

The effect of SiO_2 particles content on the electrical conductivity of PAN / SiO_2 nanocomposites is very important. In these plots, the frequency behaviour (8-12 GHz) of all nanocomposites looks like a straight line typical of hopping conduction. The absolute conductivity for individual samples increases as a function of frequency except for the nanocomposite with 15 % SiO_2. For the PAN/ SiO_2 nanocomposite the conductivity values change the order of magnitude.(10^{-5} S/cm). As the mole % of SiO_2 in PAN increases the absolute conductivity shifts to lower scales except for the 30% PAN/ SiO_2 nanocomposite. This shift is due to the electrical charges being displaced inside the nanopolymer and due to their lower concentration. The decrease in conductivity, by increase in mole % of SiO_2 may be due to particle blockage conduction path by SiO_2 (nano size particle) embedded in PAN/PU foam. Also, increase in mole% of SiO_2 leads to an increasing inter chain distance, which makes hopping between chains more difficult, resulting in reduction of conductivity. At higher frequencies the conductivities of all the samples (with different mole % SiO_2) merge to show similar frequency dependence, indicating the formation of excess charge carriers. It is clear from the figures that the PAN/ SiO_2 nanocomposite decreases the absorption percentage of aluminium plate. The results shows the variation of the microwave absorption of SiO_2 nanocomposite in PU resin matrix over the frequency range 8-12 GHz. It is interesting to note the absorption of nanocomposite over X band. Figure 4 shows the RCS diagram of nanocomposite film on 200×200 mm Aluminium plate at 8-12GHz band.

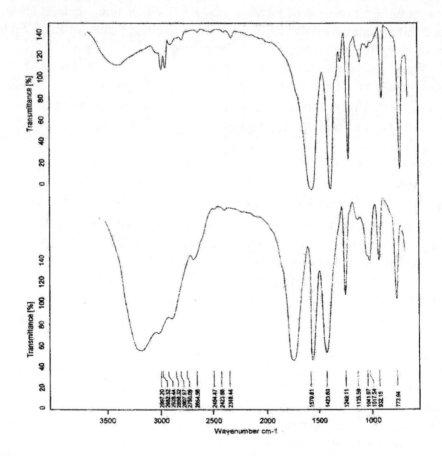

Figure 3. FT- IR spectrum of a) doped polyaniline, b) doped polyaniline in the presence of SiO_2 particles.

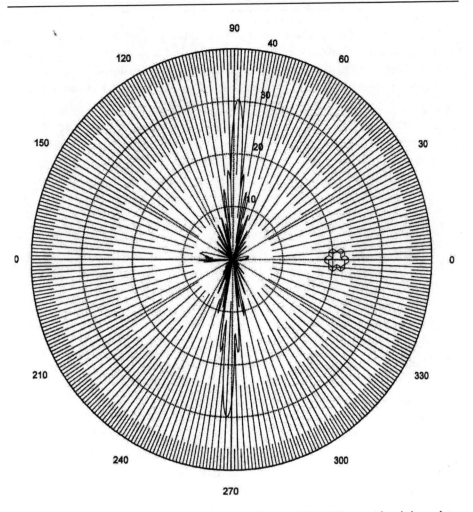

Figure 4. The RCS diagram of nanocomposite film on 200×200 mm Aluminium plate at 8-12GHz band.

CONCLUSIONS

We have synthesized a new conducting PAN foam with SiO_2 nanoparticles. PAN doping was carried out by reactive process in the presence of camphor sulfonic acid. The Silica nanoparticles(10 nm) increased electrical conductivity of foam. The conductive foam structrue have been characterized by FT-IR, SEM, and XRD techniques. This nanocomposite show crystalline nature, whereas the PAN synthezied is amorphous in nature. These nanocomposites are suitable materials for high - technology industries. The organic component is the hybride material have the dimension of 100-200 nm. One type of the composites is synthesized by preparing a precursor that contains the SiO_2 nano particles. The composites were coated on glass and metal surfaces by the method of layer-by layer coating of self – assembled multi layers. This kind of nanocomposite films have very good applications as electromagnetic shields.

REFERENCES

[1]   Ratner B.D., 1989, Comprehensive Polymer Science– The Synthesis, Characterisation, Reactions & Applications of Polymers, *Aggarwa*l, S. K., ed., vol 7, Pergamon Press, Oxford, pp. 201-247.

[2]   Saboktakin M.R., Maharramov A., Ramazanov M., A., 2007, "Synthesis and characterization of hybride polyaniline / polymethacrylic acid/ Fe_3O_4 nanocomposites ", N*ature and Science*, 5(3): 67-71.

[3]   Saboktakin M.R., Maharramov A., Ramazanov M., A., 2007, The synthesis and properties of Fe_3O_4 / Sodium acetate / CMS ternary nanocomposites as electrorheological fluid, *Journal of American Science,* 3(4):30-34.

[4]   Saboktakin M.R., Maharramov A., Ramazanov M., A., 2007, Synthesis and characterization of aromatic polyether dendrimer / Mesalamine (5-ASA) nanocomposite as drug carrier system, *Journal of American Science*, 3(4):40-45.

[5]   Li W, Tutton S, Vu AT, et al., 2005, First-Pass Contrast-Enhanced Magnetic Resonance Angiography in Humans Using Ferumoxytol, a Novel Ultrasmall Superparamagnetic Iron Oxide (USPIO)-Based Blood Pool Agent. *J Magn Reson Imaging*; 21:46-52.

[6]   Simberg D.; Duza T.; Park Ji Ho; et al; 2007, Biomimetic amplification of nanoparticle homing to tumors; *PNAS*, 2007, 104(3), 932-936.

[7]   Tang, Zh.; Alvarez N.; Yang, SZe.; 2003, Organic /Inorganic Hybride Material for Coating on Metals, *Mat.Res. Soc.Symp.Proc.*, 2003, 734, 57.1-57.8.

[8]   Berry C. C.; Curtis S.G.; 2003, Functionalisation of magnetic nanoparticles for applications in biomedicine; *J. Phys. D: Appl. Phys.*, 36, 198-206.

[9] David R.; Anne M. Wallace; Carl K. er al, 2001, A synthetic Macromolecule for sentinel Node Detection; *journal of nuclear medicine*, 42(6), 951-958.

[10] Hildebrandt, N.; Hermsdorf, D.; Signorell, A.; et al, 2007, Superparamagnetic iron oxide nanoparticles functionalized with peptides by electrostatic interactions, *ARKIVOC*, 79.

[11] Andrzej M.; Thommey P., Baker Jr., et al, 2007, Dendrimer- Based Targeted Delivery of an Apoptotic Sensor in Cancer Cells, *Biomacromolecules*, 8, 13-18.

[12] Thierry B.; Winnik F.M.; Mehri Y.; Tabrizian M.; 2003 *J.Am. Chem. Soc.* 125, 7494.

In: Modern Nanochemistry
Eds: A. K. Haghi and G. E. Zaikov

ISBN: 978-1-61209-992-7
© 2011 Nova Science Publishers, Inc.

Chapter 11

SYNTHESIS OF BIODEGRADABLE CHITOSAN NANOPARTICLES FOR BREAST CANCER THERAPY

M. R. Saboktakin[1] and A. K. Haghi[2]

[1]Department of Nanotechnology, Baku State University, Azerbaijan
[2]University of Guilan, Iran

ABSTRACT

The main objectives of our study were to prepare and evaluate a biodegradable nanoparticulate system of Letrozole (LTZ) intended for breast cancer therapy. LTZ loaded thiolated chitosan (TCS) nanoparticles were prepared by emulsion - solvent evaporation method. LTZ loaded TCS nanoaprticles were characterized by infrared spectra, drug entrapment efficiency and in vitro release. The system sustained release of LTZ significantly and further investigation could exhibit its potential usefulness in breast cancer therapy. The nanoparticles of LTZ prepared from TCS - may represent a useful approach for targeting its release at its site of absorption, sustaining its release.

INTRODUCTION

TCS which are gaining popularity because of their high mucoadhesiveness and extended drug release properties [Liu W.C.; Yao K.D.; 2002]. The derivatization of the primary amino groups of chitosan(CS) with coupling reagents bearing thiol fuctions leads to the formation of TCS [1]. The use of LTZ, which inhibits estrogen biosynthesis, is an attractive treatment for postmenopausal women with hormone – dependent breast cancer [2]. Since the early 1980s, the concept of mucoadhesion has gained considerable interest in pharmaceutical technology. If might open the door for novel, highly efficient dosage forms especially for oral drug delivery [3]. The most important goal of a cancer chemotherapy is to minimize the exposure of normal tissues to drugs while maintaining their therapeutic concentration in tomurs. Interestingly, nanoparticles (NPs) exhibits a significant tendency to accumulate in a number of tomurs after intravenous injection [4]. Hence, uptake and consequently bioavailability of the drug may be increased and frequency of dosing reduced with the result that patient compliance is improved [5, 6]. Various natural and synthetic polymers have been discovered as mucoadhesive excipients. Their mucoadhesive properties can be explained by their interaction with the glycoproteins of the mucus, based mainly on non-covalent bonds such as ionic interactions, hydrogen bonds and van der Waals forces [7, 8]. The biopolymer CS is obtained by alkaline deacetylation of chitin which one of the most abundant polysaccharides in nature [9]. Shell wastes of shrimp, lobster and crab are the main industrial sources of chitin [10]. The primary amino group accounts for the possibility of relatively easy chemical modification of CS and salt formation with acids. At acidic pH, the amino groups are protonated, which promotes solubility, whereas CS is insoluble at alkaline and neutral pH [11, 12]. Because of its favorable properties, such as enzymatic biodegradability, non-toxicity and bio-compatibility CS has received considerable attention as a novel excipient in drug delivery systems, and has been included in the European Pharmacopoeia since 2002 [13]. The administration of nanoparticles will also provide the advantage of facilitating their injection through standard infiltration needles. So far, there was one published literature on LTZ nanoparticles prepared by direct precipitation technique [14]. Recently, it has been shown that polymers with thiol groups provide much higher adhesive properties than polymers generally considered to be mucoadhesive [15]. To increase patient complicance, to overcome the undesirable side effects, LTZ could be entrapped into biodegradable nanoparticles for sustained delivery so that it can inhibit estrogen biosynthesis for a prolonged time by virtue of increased local concentration of the drug at the receptor site [16, 17]. To date, three different thiolated CS derivatives have been synthesized: CS-thioglycolic acid conjugates, CS-cysteine conjugates and chitosan-4- thio-butyl-amidine(CS-

TBA) conjugates [18]. These TCS have numerous advantageous features in comparison to unmodified CS, such as significantly improved mucoadhesive and permeation enhancing properties [19]. The strong cohesive properties of TCS make them highly suitable excipients for controlled drug release dosage forms [20]. We have prepared LTZ-loaded TCS nanoparticles (LTZ-TCS-NPs) by emulsion solvent evaporation technique to obtain smaller particle size with high entrapment efficiency and sustained release profile. Particle size, morphology, entrapment efficiency, drug – polymer interaction and *in vitro* release of LTZ- TCS-NPs were evaluated. The influence of % of drug (relation to polymer mass) on formulation performance including particle size, entrapment efficiency, *in vitro* release were investigated.

MATERIALS AND METHODS

The chemical modification of CS was performed as previously described. CS (500 mg) was dissolved in 50 mL of 1% acetic acid. In order to facilitate reaction with thioglycolic acid (TGA), 100 mg of ethyl-3-(3-dimethylaminopropyl)carbodiimide hydrochloride (EDAC) was added to the chitosan solution. After EDAC was dissolved, 30 mL of TGA was added and the pH was adjusted to 5.0 with 3 N NaOH. The reaction mixture was stirred and left for 3 h at room temperature. To eliminate the unbonded TGA and to isolate the polymer conjugates, the reaction mixture was dialyzed against 5 mM HCl five times(molecular weight cut-off 10 kDa) over a period of 3 days in the dark, then two times against 5 mM HCl containing 1.0% NaCl to reduce ionic uninteractions between the cationic polymer and the anionic sulfhydryl compound. TCS (0.2 g) was dissolved in 15 mL 1% vol/vol acetic acid containing 4% glycerine. For the loading of LTZ into polymeric matrix, 0.24 g LTZ was suspended in the mixture. Before the addition of nanoparticles, a 200 μl sample was taken and filtered using a low protein binding 0.22 μm PVDF filter (Millipore, Bedford, MA) and then replaced with equal amounts of 1 × PBS, pH 7.4. An additional sample was taken in the same manner after loading. The particles were collaped using 10 mL 0.1 N HCl, filtered with Whatman Grade 4 filter paper, and washed with 20 mL of deionized water. After filtering, particles were frozen in a -80°C freezer and lyophilized at -50°C under vacuum (LabConco Model 77500) for 24 hours. TCS – LTZ nanoparticles were prepared as previously described and kept in a dry environment until imaging.

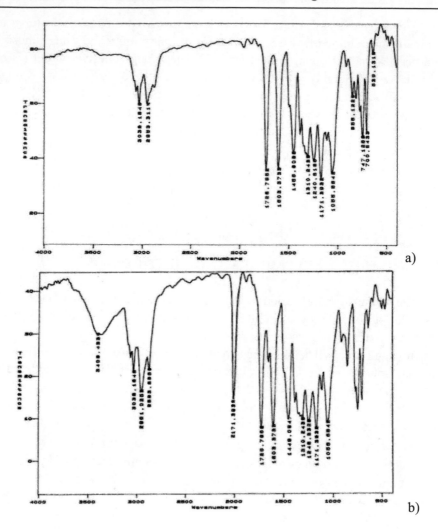

a)

b)

Figure 1. FT-IR spectra of a) pure thiolated chitosan b)Letrozloe loaded thiolated chitosan(TCS) nanoparticles.

The nanoparticles were sprinkled onto an aluminum stub that was covered with carbon tape. Excess nanoparticles were removed by gently tapping the stub and the samples were sputtered coated with a gold layer between 5 and 10 nm thick. Samples were imaged with a SEM (Philips XL-30 E) at 10kV and a working distance of 7 mm. The Fourier transform infrared analysis was conducted to verify the possibility of chemical bonds between drug and polymer. Samples of pure LTZ and LTZ-TCS nanopartilces were scanned in the IR range from 400-4000 cm^{-1} with carbon black as reference. The detector was purged carefully by clean dry helium gas to increase the signal level and reduce moisture. Drug entrapment efficiency (EE) was determined by centrifuging (with Centrifuge-3K30, Sigma Laboratory) aqueous dispersion of LTZ-TCS nanoparticles at 25, 000 rpm, 5°C for 25 min and measuring the amount of LTZ in the supernatant with the help of double beam UV-VIS Spectrophotometer (Philips PU 8620, USA), set at 238 nm. The amount of LTZ was subtracted from initial amount of LTZ taken to calculate drug entrapment efficiency of nanoparticles. The experiment was performed in triplicate for each batch and average drug entrapment efficiency was calculated. In vitro release study of LTZ-TCS nanoparticles was conducted in Franz Diffusion Cell. The diffusion cell model adapted to the spectrophotometer cuvette, with 1 cm of optic way and 1 mL of volume, was used for the *in vitro* release of LTZ. A cellulose acetate membrane (Dialysis membrane with Molecular weight cut off value of 5, 000 – 10, 000, Himedia-60) was adapted to the terminal portion of the glass cylinder of the Franz Diffusion cell by a rubber ring. mL of LTZ-TCS nanoparticles aqueous dispersion was loaded to cylinder and coupled to the diffusion cell containing the receptor phase(60 mL of 0.2 M Phosphate buffer solusion pH 7.4) at 37°C. The dissolution media was agitated at 25 rpm using magnetic stirrer. At different time intervals, aliquots of 2 mL were withdrawn and immediately restored with same volume of fresh Phosphate buffer. The amount of LTZ released was assessed by double beam UV spectrophotometer (Philips PU 8620, USA) set at 238 nm *versus* a calibration curve prepared in the same buffer.

RESULTS AND DISCUSSION

The synthesized polymer was characterized by IR spectroscopy. The FT-IR spectrum of LTZ- TCS is shown in Figure 1. In the spectra of TCS –OH and –NH stretch were clearly seen at 3038.185 and 2961.025 cm^{-1}, respectively. Additional presence of amidine I and amidine II bands are seen at 1725.786 and 1603. 373cm^{-1} corresponding to $\geq NH_2^+$ stretch and $NH \geq NH_2^+$, respectively, being two coupled vibrations. The presence of an additional band at 1495.3783 and 1449.0847 cm^{-1} can be assigned for N-H band of the salt $NH_2^+Cl^-$. Other characteristic peaks of CS O-H stretch, C-H stretch and C-O stretch were present at 3400-3600, 2930, and 1009-1171 cm^{-1}, respectively. This confirmed the synthesis of TCS. The spectrum of TCS was well correlated with reports by Matsuda *et al.* for TCS. Thiol content of TCS was found 214±52 μmol/g.

The morphology of nanoparticles was examined by scanning electron microscopy (SEM, Philips XL-30 E, USA). The nanoparticles were mounted on metal stubs using double – sided tape and coated with a 150 Å layer of gold under vacuum. S tubes were visualized under scanning electron microscope. Figure 2 shows the SEM of LTZ-TCS nanoparticles that synthesized by chemical reaction. This nanoparticles is very sensitive to the temperature that due to the interaction electron and sample. Scanning electron micrography images were obtained from a diluted solution of the LTZ particle. The white spots are LTZ nanoparticles. The SEM image shows the presence of LTZ spherical particles in polymer matrix, which are homogenenously distributed throughout the polymer, which is also confirmed from ^{1}H-NMR studies. As observed from SEM photomicrographs, the crystals of LTZ have a different appearance than recrystallized LTZ.

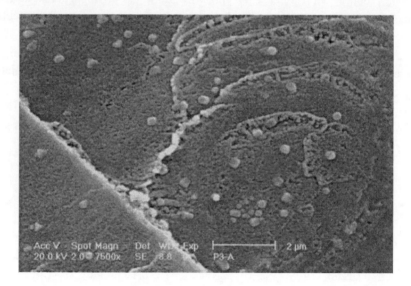

Figure 2.

These nanoparticles do not have clearly defined crystal morphological features in the SEM photomicrographs. It has been reported that the encapsulation efficiency of LTZ-TCS nanoparticles increases from about 68.5 % to 87.2% with the increment of their mean diameter from 64 nm to 255 nm (Table 1). The lower encapsulation efficiencies obtained with the smaller particles could be explained by the longer surface area of smaller droplets for a given volume of organic phase. Hence, during the emulsification step, a more direct contact between internal and external phases occurred, resulting in a higher drug loss by diffusion towards the external medium.

Table 1. Characterization of blank and LTZ-TCS nanoparticles

Formulation Drug (wt% of polymer)	Encapsulation Efficiency(%)	Mean diameter (nm)	Poly-dispersity index
Blank	–	60.4±25.0	0.27±0.01
10	68.50±0.52	64.0±15.0	0.36±0.05
20	76.12±3.6	145.9±40.2	0.45±0.01
30	87.20±4.0	255.3±40.5	0.68±0.10

Batches of three different compositions (10%, 20%, 30% of LTZ) were studied and the results of cumulative percentage released over 15 hours were shown in Figure 3.

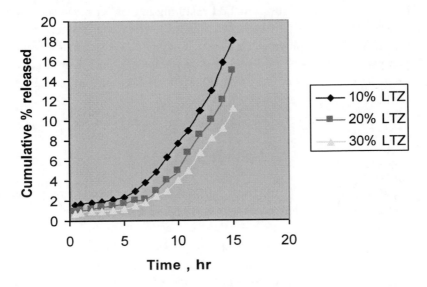

Figure 3. In vitro release data of LTZ-TCS nanoparticles.

The results showed that there was a pronounced time prolongation of drug release from LTZ-TCS nanoparticles. Batch with 10% LTZ showed the slowest release and batch with 30% LTZ showed highest release. Formulation(30% drug) with highest LTZ entrapment (86.32%) showed highest drug release rate and cumulative drug release while formulation with

10% drug and lowest LTZ entrapment increases the amount of drug close to the surface as well as the drug in the core from nanoparticles. Thus the difference in drug release profiles of three batches of LTZ-TCS nanoparticles can be explained. Drug release from LTZ-TCS nanoparticles takes place by several mechanisms including surface and bulk erosion, distintegration, diffusion, and desorption. In this study, release of drug from TCS matrix has been found to occur predominantly by its diffusion from the polymer matrix. During the later phases, the release is mediated through both diffusion of therapeutic agent and degradation of polymer matrix itself.

CONCLUSIONS

The results of present study revealed that LTZ can be entrapped into TCS nanoparticles, which can provide sustained drug release with high drug entrapment efficiency. The nanoparticles were successfully developed by emulsion / solvent evaporation method resulting in smaller mean particle size range. Size distribution, entrapment efficiency, release characteristics were influenced by drug to polymer ratio in formulation. Hence, nanoparticles of LTZ prepared from TCS may represent a useful approach for targeting its release at its site of absorption, sustaining its release and improving its oral availability.

REFERENCES

[1]   Liu W.C.; Yao K.D.; (2002); "Chitosan and its derivaties-a promising non-viral vector for gene transfection"; *Journal of Control Release*; 83:1-11.

[2]   Synder G.H.; Ready M.K.; Cennerazzo M.J.; Field D.; (1983); "Use of local electrostatic environments of cysteines to enhance formation of a desired species in a reversible disulfide exchange reaction"; *Journal of Biochemical Biophysical Acta*; 749: 219-226.

[3]   Thanou M.; Nihot M.T.; Jansen M.; Verhoef J.C.; Junginger J.C.; (2001); "Mono-N-carboxymethyl chitosan (MCC), a polyampholytic chitosan derivative, enhances the intestinal absorption of low molecular weight heparin across intestinal epithelia in vitro and in vivo "; *Journal of Pharmaceutical Sciences*; 90:38-46.

[4]   Geisler J.; Helle H.; Ekse K.; Duong N.K.; Evans D.B.; Nordbo T.; Lonning P.E.; (2008); "Letrozole is superior to anastrozole in suppressing breaset cancer tissue and plasma estrogen levels "; *Journal of Clinical Cancer Research*; 14(19); 6330-6335.

[5]   Bibby D.C.; Talmadge J.E.; Dalal M.K.; Kurz S.G.; Chytil K.M.; Barry S.E.; Shand D.G.; Steiert M.; (2005); "Pharmacokenetics and biodistribution of RGD-targeted doxorubicin loaded nanoparticles in tumor –bearing mice "; *International Journal of Pharmaceutics;* 293(1-2); 281-290.

[6]   Andreas B.S.; Hopf T.E.; (2001); "Synthesis and in vitro evaluation of chitosan-thioglycolic acid conjugates "; *Journal of Science pharmacy*; 69: 109-118.

[7]   Hornof M.D.; Kast C.E.; Andreas B.S.; (2003); "In vitro evaluation of the viscoelastic behavior of chitosan-thioglycolic acid conjugates "; *European Journal of Pharmaceutical Biopharmaceutics*; 55:185-190.

[8]   Andreas B.S.; Hornof M.; Zoidl T.; (2003); "Thiolated polymers-thiomers: modification of chitosan with 2-iminothiolane "; *International Journal of Pharmaceutics;* 260: 229-237.

[9]   Roldo M.; Hornof M.; Caliceti P.; Andreas B.S.; (2004); Mucoadhesive thiolated chitosans as platforms for oral controlled drug delivery: synthesis and in vitro evaluation "; *European Journal of Pharmaceutical Biophamaceutics*; 57(1): 115-121.

[10]   Langoth N.; Guggi D.; Pinter Y.; Andreas B.S.; (2004); "Thiolated chitosan: in vitro evaluation of its permeation properties "; *Journal of Control Release*; 94(1): 177-186.

[11]   Kast C.E.; Valenta C.; Leopold M.; Andreas B.S.; (2002); "Design and in vitro evaluation of a novel bioadhesive vaginal drug delivery system for clotrimazole"; *Journal of Control Release*; 81:347-354.

[12]   Leitner V.M.; Marschutz M.K.; Andreas B.S.; (2003); "Mucoadhesive and cohesive properties of poly(acrylic acid)-cysteine conjugates with regard to their molecular mass"; *European Journal of Pharmaceutical Sciences*; 18:89-96.

[13]   Senel S.; Kremer M.; Kas S.; Wertz P.W.; Hincal A.A.; Squier C.A.; (2000); "Enhancing effect of chitosan on peptide drug delivery across buccal mucosa; *Journal of Biomaterials;* 21: 2067-2071.

[14]   Mondal N.; Pal T.K.; Ghosal S.K.; (2008); "Development, physical characterization, micromeritics and in vitro release kinetics of letrozole loaded biodegradable nanoparticles"; *Pharmazie*; 63(5); 361-365.

[15] Felt O.; Buri P.; Gurny R.; (1998); "Chitosan: a unique polysaccharide for drug delivery"; *Journal of Drug Development and Industrial Pharmacy*; 24:979-993.

[16] Kast C.E.; Frick W.; Losert U.; Andreas B.S.; (2003); "Chitosan – thioglycolic acid conjugate: a new scaffold material for tissue engineering "; *International Journal of Pharmaceutics;* 256: 183-186.

[17] Leroux J.C.; Allemann E.; De Jaeghere F.; Doelker E.; Gurny R.; (1996); "Biodegradable nanoparticles from sustained release formulations to improved site specific drug delivery "; *Journal of Control Release*; 39; 339-350.

[18] Coppi G.; Lannuccelli V.; Leo E.; Bernabei M.T.; Cameroni R.; (2001); "Chitisan-alginate microparticles as a protein carrier "; *Journal of Drug Development and Industrial Pharmacy*; 27: 393-400.

[19] Guggi D.; Kast C.E.; Andreas B.S.; (2003); "In vivo evaluation of an oral calcitonin delivery system for rats based on a thiolated chitosan matrix; *Journal of Pharmaceutical Research*; 20(12): 1989-1994.

[20] Bromberg L.E.; (2001); "Enhanced nasal retention of hydrophobically modified polyelectrolytes "; *Journal of Pharmacy and Pharmacology*; 53: 109-114.

In: Modern Nanochemistry
Eds: A. K. Haghi and G. E. Zaikov

ISBN: 978-1-61209-992-7
© 2011 Nova Science Publishers, Inc.

Chapter 12

SYNTHESIS OF WATER-DISPERSIBLE NANOCOMPOSITES AS COATING ON METALS

M. R. Saboktakin[1] and A. K. Haghi[2]
[1]Department of Nanotechnology, Baku State University, Azerbaijan
[2]University of Guilan, Iran

ABSTRACT

The effect of selenium oxide (SeO_2) nanoparticles addition on the physicochemical properties of the polyaniline (PAN) / polymethacrylic acid (PMAA) was investigated. In the presence of SeO_2 nanoparticles, PAN/PMAA was observed in the form of discrete nanoparticles, not the usual network structure. PAN/PMAA showed crystalline structure in the nanocomposites and pure PAN/PMAA formed without SeO_2 nanoparticles. PAN/PMAA exhibited amorphous structure and nanoparticles were completely etched away in the nanocomposites formed with the mechanical stirring over a 7-h reaction. The electrical conductivity of the nanocomposites increased greatly upon the initial addition of SeO_2 nanoparticles. Standard four-probe measurements indicated a three-dimensional variable-range-hopping conductivity mechanism.

INTRODUCTION

Polymer nanocomposites constitute a class of hybrid materials composed of a polymer matrix and an inorganic component which has at least one dimension in the nanometer (<100 nm) size domain. Nanoparticles were first developed around 1970. The synthesis of metallic nanoparticles has become an extremely interesting topic in the field of material science, due to the wide range of optical and electronic properties that are accessible in the nanometer – size regime. Nanometer scale hybride materials is currently an area of active research. [1]. These nanoparticles are hybride between organic and transition metals or rare earth oxides, for example cerium oxide, ferrous oxide, ... [2]. metallic nanocomposites have emerged as a new class of materials because of their unique electrical, optical and chemical properties [3]. The properties of a polymer – reinforced composite are mostly influenced by the size, shape, composition, state of agglomeration, and degree of matrix inorganic component [4]. Decreasing the particle size to the nano-size dimension influence the macroscopic properties of the polymer because a breakdown of the common rule of mixture theory occurs [5]. This breakdown is caused by the amount of interfacial zone that gains importance with respect to the phase relative to bulk behavior [6].

Demonstrated application of such composites can be found, among others, in the fields of optics, mechanics, iono-electronics, biosensors, flame retardants and membranes. Polyaniline, alow cost intrinsically conductive polymer, is stable under ambient conditions [7]. Because of it is low solubility in most common solvents, the industrial application of PAN is limited. For the purpose of improving it's solubility in water and to prevent dopant migration, PAN has been modified in the self-doped form by introducing protonic acids into the side chains [8]. For example, a sulfonic acid group was introduced into the benzenoid ring to give sulfonated PAN obtained by reacting the emeraldine base with fuming sulfonic acid, an sulfonic acid group or an alkylbenzene sulfonic acid group was grafted on amine nitrogen to give a water – dispersible polymers [9, 10]. Cerium oxide and PAN have been studied for its electroactive interaction with metal surfaces for protecting the lightweight high - strength aluminum alloys from corrosion [11]. The research in this area has been motivated by the need for a new corrosion inhibitor that may replace the toxic chromates currently in use. Since the inorganic oxide and the conducting polymer protect the metal surface by different mechanisms [12, 13], there is a possibility for the hybrid materials to be a more effective corrosion inhibitor due to the synergistic interaction provided by the hybride

[14, 15]. In this paper, the effect of SeO_2 nanoparticle addition on the morphology of PAN/PMAA and electrical conductivity of resulting SeO_2/PAN-PMAA nanocomposites was reported. The effect of the stirring method, i.e., ultrasonic and mechanical stirring on the composite fabrication was also reported. The electrical conductivity was investigated by a standard four-probe method and found to be strongly dependent on the particle loadings. The SeO_2 nanoparticles were observed to be stable even after exposure to a strong acid with a pH value of 1.0 for more than 3 weeks.

MATERIALS AND METHODS

0.012 mole (3.5 gr) PMAA ($M_w = 90,000$) dissolve in 25 grams distilled water. Stir for 2 hours. Then add 0.558 gram of aniline to solution, stir for 4 hours to allow equilibrium adsorption. The molar ratio of the PMAA component to the aniline is 2:1. Acidifiy the solution prepared by adding 4 ml 3M Nitric acid. Dissolve 0006 mol ammonium persulfate with 10 ml distilled water. Mix these two solutions to start the polymerization reaction. The solution turned to dark green. Stir for 24 hours to obtain a homogeneous solution of the polymeric complex. The polymeric complex is a dispersion of particle of 100-200 nm in diameter. The infrared absorption spectra of the complex are consistent with the structure of a polymeric complex of PAN and PMAA. Stir SeO_2 dispersive solution with sodium dodecyl sulfate / distilled water with mechanical and ultrasonic stirring were explored, and the resulting nanocomposite properties were characterized accordingly. A diluted solution of the PMAA: PAN complex is then mixed with SeO_2 solution (10% SeO_2) for 7-h. The average molar ratio of the components is PMAA:PAN: $SeO_2 = 2:1:2$. The dispersion is stable with very small amount of precipitation. All the products were washed thoroughly with deionized water (to remove any oligomers), respectively. The precipitated powder was dried at 50°C for further analysis. As a control experimental for comparison purposes, pure PAN/PMAA was also synthesized following the same procedure as described before but without SeO_2 nanoparticles.

RESULTS AND DISCUSSION

Figure 1 shows the SEM nanostructures of the synthesized SeO_2/ PAN-PMAA nanocomposites fabricated by the conventional method, as used for SeO_2 nanoparticles filled PAN/PMAA composite fabrication. The conventional method is based on mechanical stirring over a long period of time. In contrast to the network structure of pure PAN/PMAA as shown in Figure 1, discrete spherical nanoparticles with uniform size distribution are observed in the nanocomposite counterparts fabricated by the conventional method. However, no attraction was observed when a permanent magnet was placed nearby, and further quantitative magnetic characterization did not show any sign of magnetization in the nanocomposite. The disappearance of the SeO_2 nanoparticles is believed to be due to the slow dissolution over time caused by acidic solution used in the aniline polymerization. This observation also suggests the formation of a porous PAN/PMAA shell rather than a solid tight one that can protect the SeO_2 nanoparticles from dissolution.

The crystallinity of the formed composites was followed with X-Ray diffraction (XRD) as s function of weight percent inorganic component. Studies on XRD patterns of PAN/PMAA are scarce in the literature. Figure 2 shows the XRD pattern for PAN-PMAA- SeO_2 (10%). The diffraction pattern of PAN-PMAA- SeO_2 nanocomposite shows a peak at about $2\Theta = 26.16°$.

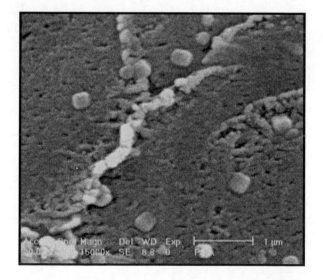

Figure 1. Scanning electron micrograph of PMAA/PAN/SeO_2 nanocomposite.

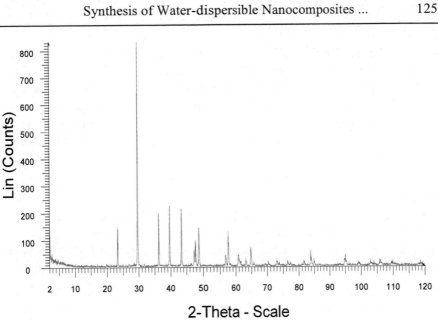

Figure 2. XRD spectra of PAN/PMAA/ SeO$_2$ nanocomposite.

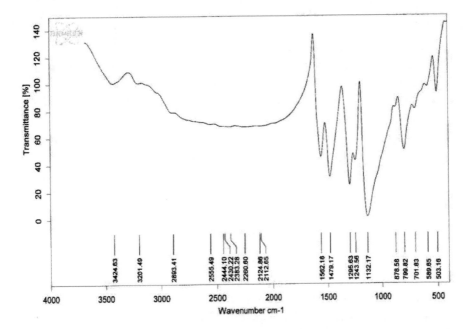

Figure 3. FT-IR spectra of PAN/PMAA / SeO$_2$ nanocomposite.

Figure 3 shows the FT-IR spectrum of PAN/PMAA nanocomposite, where the % of transmittance is plotted as a function of wave number (cm^{-1}). The characteristic FT-IR peak at 1523 and 1485 cm^{-1} are due to the presence of quinoid and benzenoid rings, respectively and are clear indication of these two states in the polymer chain. Also, The peaks at 1176, 1710 cm^{-1} are due to the C-N, C=O bond stretching vibration, respectively. The FTIR spectra of PAN/PMMA composite in presence of SeO$_2$ exhibite new adsoption peaks distinctly at 1562, 1479, 1295, 1132 and 799 cm^{-1} which are assignable to the presence of various metal oxide in the composite. The obvious spectral differences between pure PAN/PMAA and the composites indicate the PAN/PMAA exhibits a different PAN/PMAA chain structure and there are physicochemical interactions between the nanoparticles and PAN/PMAA. The presence of SeO$_2$ nanoparticles is strongly supported by new peaks 503 cm^{-1} and 580 cm^{-1} as shown in Figure 3, which are characteristic peaks of SeO$_2$. This observation indicates that a conductive nanocomposite can be synthesized with PAN/PMAA if polymerization can be achieved in a short period of time. The prolonged polymerization is characterized by the disappearance of the characteristic IR peaks of SeO$_2$. The almost similar spectra between the PAN/PMAA and PAN/PMAA fabricated by the long time reaction also indicate the loss of SeO$_2$ nanoparticles.

The organic / inorganic nanocomposites are stable dispersion in water. We have exploied the possibility for coating glass and metal substrates with the nanocomposite.There are several different methods for coating the material on substrates. The PAN/PMAA /SeO$_2$ particles have a negatively charged surface due to the partially ionized carboxylic functional group. The anionic particle is then adsorbed on the positively charged poly(ethyleneimine) layer. At this point, the first bi-layer structure is formed on the substrate. At this process is repeated, we build up a structure of coating consisting repeated bi-layers. The green color of the coating becomes more apparent as the number of bilayers is increased. Figure 4 shows the optical absorption at 800nm (the polaron absorption band of polyaniline) of the self – assembled bilayers coated on a glass substrate.

As more layers of the composite material are coated, the UV – visible absorption spectra increase in intensity (Figure 5). The linear relationship between the absorbance and the number of bilayers indicated the thickness of the bilayers is constant which is consistent with the molecular assembly nature of the coating.

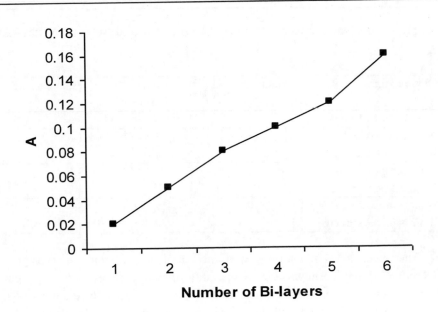

Figure 4. Absorbance as a function of the number of bilayers.

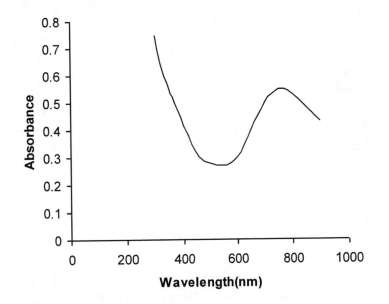

Figure 5. UV-Vis absorption spectra of PAN/PMAA / SeO_2 nanocomposite.

**Table 1. Physical properties of pure PAN/PMAA and SeO₂/PAN-PMAA
nanocomposites**

Material name	Conductivity at 290K (S.cm^{-1})	Conductivity at 10K (S.cm^{-1})	Final particle loading (wt%)
Pure PAN/PMAA	0.9	1.5×10^{-4}	0
Nanocomposite 10% (1 h ultrasonic stirring)	7.6	112.6×10^{-4}	1.6
Nanocomposite 10% (7 h ultrasonic stirring)	1.6	8.6×10^{-4}	8.6
Nanocomposite 10% (7 h mechanical stirring)	$< 1.5 \times 10^{-9}$	$< 1.5 \times 10^{-9}$	5.6

The electrical conductivity was measured by a standard four-probe
method. Figure 6 shows the temperature-dependent resistivity of pure
PAN/PMAA and nanocomposites with different particle loadings. The
PAN/PMAA synthesized with the presence of the nanoparticles over a long-
time reaction was observed to have large resistance beyond the ability of the
utilized equipment due to the poor contact between small particles and the
observed amorphous structure. At lower temperature the resistance is too large
to be measured by the utilized equipment. The much lower resistivity in the
pure PAN/PMAA prepared by the conventional method is due to the network
structure formation as shown in the SEM image of Figure 1, which favors
electron transport.

The resistivity change is not due to the doping extent based on the fact that
all the samples were washed with the same solution. The resistance was
observed to decrease dramatically in the SeO₂ /PAN-PMAA nanocomposites.
The nanocomposite with an initial particle loading of 10 wt% was observed to
have a higher resistivity, which is due to the known insulating properties of
SeO₂. The effect of the stirring method, i.e., mechanical versus ultrasonic
stirring was investigated by polymerizing for 7 h as used in the mechanical
stirring. Figure 7 shows the hysteresis loop of the as-received SeO₂
nanoparticles and the nanocomposite (an initial particle loading of 10%)
synthesized with ultrasonic stirring over 7 h, respectively. The weight
percentage of SeO$_2$ nanoparicles in the nanocomposite was estimated to be
5.6% based on the nanocomposite and as-recieved nanoparticles.

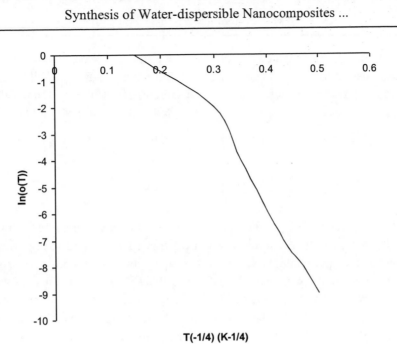

Figure 6. Temperature-dependent conductivity of nanocomposite.

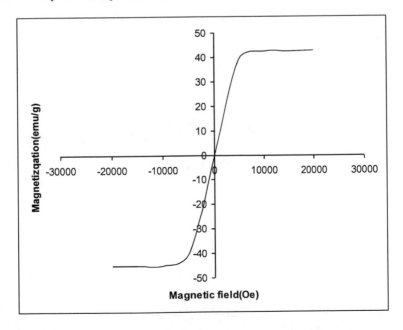

Figure 7. Magnetic hysteresis loops of nanocomposites.

This weight percentage is much lower than the initial particle loading and the composite sample synthesized with a 1-h ultrasonic stirring. This indicates that more particles are lost due to the dissolution over the protons. Table 1 shows the physical properties of pure PAN/PMAA and SeO$_2$/PAN-PMAA nanocomposites.

CONCLUSIONS

The effect of SeO$_2$ nanoparticles on the chemical polymerization of PAN/PMAA in an acidic solution was investigated and found to significantly influence the morphology (size and shape) and other physicochemical properties of the PAN/PMAA. Pure discrete PAN/PMAA nanoparticles with a much higher resistivity are formed over a long reaction time in the presence of SeO$_2$ nanoparticles. Similar to pure PAN/PMAA formed with SeO$_2$ nanoparticles and different from the network structure of the pure PAN/PMAA formed without SeO$_2$ nanoparticles, discrete nanoparticles are observed in all the nanocomposites with an initial particle loading of 10 wt%. The decreased conductivity in the high particle loading is due to the insulating behavior of SeO$_2$ nanoparticles. The nanocomposites were coated on glass and metal surfaces by the method of layer-by-layer coating of self-assembled multi layers.

REFERENCES

[1]   Ratner B.D., 1989, *Comprehensive Polymer Science– The Synthesis, Characterisation, Reactions & Applications of Polymers,* Aggarwal, S. K., ed., vol 7, Pergamon Press, Oxford, pp. 201-247.

[2]   Saboktakin M.R., Maharramov A., Ramazanov M., A., 2007, Synthesis and characterization of hybride polyaniline / polymethacrylic acid/ Fe$_3$O$_4$ nanocomposites *Nature and Science*, 5(3): 67-71.

[3]   Saboktakin M.R., Maharramov A., Ramazanov M., A., 2007, The synthesis and properties of Fe$_3$O$_4$ / Sodium acetate / CMS ternary nanocomposites as electrorheological fluid, *Journal of American Science*, 3(4):30-34.

[4]   Saboktakin M.R., Maharramov A., Ramazanov M., A., 2007, Synthesis and characterization of aromatic polyether dendrimer / Mesalamine (5-ASA) nanocomposite as drug carrier system, *Journal of American Science*, 3(4):40-45.

[5]   Li W, Tutton S, Vu AT, et al., 2005, First-Pass Contrast-Enhanced Magnetic Resonance Angiography in Humans Using Ferumoxytol, a Novel Ultrasmall Superparamagnetic Iron Oxide (USPIO)-Based Blood Pool Agent. J Magn Reson Imaging; 21:46-52.

[6]   Simberg D.; Duza T.; Park Ji Ho; et al; 2007, Biomimetic amplification of nanoparticle homing to tumors; *PNAS*, 2007, 104(3), 932-936.

[7]   Berry C. C.; Curtis S.G.; 2003, Functionalisation of magnetic nanoparticles for applications in biomedicine; *J. Phys. D: Appl. Phys.*, 36, 198-206.

[8]   David R.; Anne M. Wallace; Carl K. er al, 2001, A synthetic Macromolecule for sentinel Node Detection; *journal of nuclear medicine*, 42(6), 951-958.

[9]   Hildebrandt, N.; Hermsdorf, D.; Signorell, A.; et al, 2007, Superparamagnetic iron oxide nanoparticles functionalized with peptides by electrostatic interactions, *ARKIVOC*, 79.

[10]   Andrzej M.; Thommey P., Baker Jr., et al, 2007, Dendrimer- Based Targeted Delivery of an Apoptotic Sensor in Cancer Cells, *Biomacromolecules*, 8, 13-18.

[11]   Thierry B.; Winnik F.M.; Mehri Y.; Tabrizian M.; 2003*J.Am. Chem. Soc.*, 125, 7494.

[12]   Stoffer, J.O., O'Keefe, T.J., Lin, E., Morris, Yu, S.P., Sitaram, S.P., US Patent 5, 932, 083.

[13]   Racicot, R., Brown, R., Yang, S.C., "Surface conversions of Alumium and ferrous Alloys for corrosion Resistance", *Proceedings of corrosion*, 2000, 113-128.

[14]   McCarthy, P.A., Huang, J., Yang, H.L., *Lagmuir*, 2002, 18, 259.

[15]   Cheung, J.H., Stockton, W.B., Rubner, M.F., *Macromolecules*, 1997, 30, 2712.

In: Modern Nanochemistry
Eds: A. K. Haghi and G. E. Zaikov

ISBN: 978-1-61209-992-7
© 2011 Nova Science Publishers, Inc.

Chapter 13

KINETIC STUDIES OF HYDROGENATION OF 2, 4-DINITROTOLUENE OVER RANEY-NICKEL NANOPARTICLES

M. R. Saboktakin[1] and A. K. Haghi[2]

[1]Department of Nanotechnology, Baku State University, Azerbaijan
[2]University of Guilan, Iran

ABSTRACT

The dependence of 2, 4-dinitrotoluene(2, 4-DNT) hydrogenation rate upon catalyst particles size and kinetic parameters in ethanol-water-2, 4-DNT three components solvent over Raney-Ni nanosized catalyst (20nm) has been studied. More stringent requirements for reducing the content of aromatics have brought increasing attention to new catalysts involved in hydrotreating process. The investigation of the effect of particles size of catalyst on the rate of the reaction where done in the region is corresponding to the zero reaction order. An optimum set of conditions for reaction occurrence is established. To confirm the absence of gas-liquid, liquid-solid, and intraparticle mass transfer effects on the reaction, the effects of stirring speed, catalyst loading, and catalyst particle size on the initial reaction rate at the maximum temperature and 2, 4- DNT concentration have been thoroughly studied. The results show that for a catalyst particle size and constant stirring speed, the reaction rate is not influenced by the mass-transfer processes. Also, effective intraparticle diffusivity of 2, 4-DNT has been determined from the effectiveness factor of the catalyst for its different particle sizes. The main objective of this paper was to investigate the catalytic hydrogenation of 2, 4-DNT for the preparation of 2, 4-DAT and the fundamental issues such as adsorption and reaction kinetics relevant to these processes.

INTRODUCTION

Synthesis of 2, 4-diaminotoluene (2, 4-DAT) by catalytic hydrogenation of 2, 4-DNT and compounds containing nitro group with different subtitueants was investigated in liquid phase and gas phase over different types of catalysts with using a solvents of different nature and composition [1]. A nitro group catalytic hydrogenation in substituted nitrobenzenes is the base for synthesizing a number of products of thin organic synthesis, aromatic amines in particular. It is well known that nitrobenzene hydrogenation product-aniline is widely used in synthesis of various dyes [2-4]. The rate and the selectivity of aromatic nitro-compounds hydrogenation depend upon different factors such as temperature, hydrogen pressure and concentrations of catalyst and a hydrated compound. It is known that if hydrogen pressure is increased hydrogenation rates of nitro compounds is increased well [5, 6]. When hydrogen adsorption is not a limiting stag of the process and the reaction proceeds at low concentration of a hydrated compound the nature of reaction rate dependence upon the pressure of hydrogen will be determined by the isotherm form of hydrogen adsorption. The degree of diffusion retarding influence significantly the character of hydrogenation rate-hydrogen pressure dependence [7, 8]. The catalytic hydrogenation of nitrobenzene to p-aminophenol(PAP) in a four – phase system is a unique example of a multiphase catalytic reaction system, in which the intermediate (phenylhydroxyamine) formed in the organic phase is first transferred to the immiscible aqueous acidic phase where it undergoes a acid catalyzed rearrangement to give the desired final product (PAP).While this route has been shown to be feasible with reasonably high selectivity to the PAP, there are several issues, which need further investigations [9, 10]. This system involves four different phases namely hydrogen as a gas phase, Raney-Nickel catalyst as a solid phase, nitrobenzene as an organic phase and aqueous sulfuric acid as the second immiscible liquid phase [11-13]. Such a complex system poses a series challenge in operation on a commercial scale, as numerous uncertainties in scale-up of such multiphase system are likely to exist [14]. Like in any heterogeneous catalytic reaction, the adsorption characteristics of various reaction components on the catalyst surface would also have a significant impact on the overall performance of the catalyst as well as the rate of reaction [15, 16]. The main purpose of this study was to evaluated the effects of hydrogen pressure, temperature and other parameters on the rate of 2, 4-DNT hydrogenation over Raney-Nickel nanoparticles taking into account the diffusion retarding kinetic region and adsorption factors in ethanol-water two components solvent [17, 18]. Like in any heterogeneous catalytic reaction, the adsorption characteristics of various reaction components on the overall performance of the catalyst as well as the rate of reaction. Therefore, detailed investigations of adsorption of 2, 4-DNT, 2, 4-DAT and hydrogen on Raney-Ni catalyst was also undertaken. The catalytic hydrogenation rate and average catalytic activity was undertaken to select the optimum reaction conditions. Further, a detailed kinetic study of hydrogenation of 2, 4-DNT to 2, 4-DAT was performed to purpose a rate equation to represent the kinetic date satisfactorily [19, 20].

2. MATERIALS AND METHODS

2.1. Materials

2, 4-DNT, Raney-Ni nanopartilces (20 nm) was prepared by Nanotechnology Research Center of Baku State University. Raney-Ni in micro scale was purchased from Aldrich Chemicals. Hydrogen gas was supplied by Azar Oxide Iran Ltd. Solvents were obtained from Merck Chemicals, HPLC grade water and acetonitrile (Merck Chemicals, Germany) were used as received.

2.2. Analytical Measurements

The analysis of all liquid samples was carried out using a Hewltt-Pachard gas Chromatograph model 6890 equipped with HP-1 capillary column(30m length, 0.32 mm diameter) and flame ionization detector (FID). The analysis conditions were; column temperature: 433K, injection temperature: 523K, detector temperature(FID): 523K, carrier gas(nitrogen) flow rate = 30ml/min. The quantitative analytical procedure had relative accuracy within 2%.Calibration factors were determined by using liquid standards having known compositions of 2, 4-DNT and 2, 4-DAT. The water content in ethanol was measured by Karl-Fischer method.

2.3. Experimental Procedure

In a typical hydrogenation experiment, Raney-Ni nanopartilces (8mg, $0.2660 g/cm^3$), 2, 4-DNT (2mg, $0.4792 mol/cm^3$) and 30 ml ethanol(90%) were charged into a clean autoclave. The contents were flushed with nitrogen to ensure removal of traces of oxygen from the reactor. The reactor was then heated to 353K under slow stirring (100 rpm) and the temperature was allowed to stabilize at the desired set point. Then, hydrogen gas was introduced to a desired level (1 atm) and the contents stirred vigorously (1000 rpm). The reactor was operated at a constant pressure throughout the reaction period by supplying hydrogen from a reservoir vessel was measured as a function of time using a digital pressure transducer. After completion of the reaction, the reactor was cooled and the excess gas was vented off. The product 2, 4-DAT was separated as a solid product in the autoclave due to its low solubility in the reaction medium at room temperature. The reaction mass was further diluted to 60 ml by acetonitrile and the resulting solution was analyzed on GC for 2, 4-DNT and 2, 4-DAT.

3. RESULTS AND DISCUSSION

Figure 1 shows that on the kinetic curves two characteristic plots can be underlined. The first plot corresponds to the constant 2, 4-DNT hydrogenation rate and the second one corresponds to the directly proportional change of hydrogenation rate upon the 2, 4-DNT concentration. The effect of 2, 4-DNT concentration on the initial rate of hydrogenation and activity of the catalyst was studies. It was observed that the initial rate of hydrogenation decreased.The activity of the catalyst was found to increase with 2, 4-DNT concentration.

Constant 2, 4-DNT hydrogenation rate corresponds to the zero reaction order with respect to the hydrated compound and its decrease corresponds to the first order reaction with respect to 2, 4-DNT. The region of the changing of the reaction order with respect to 2, 4-DNT corresponds exactly to the region of deviation from the linear dependence of 2, 4-DNT concentration upon the time(Figure 2).

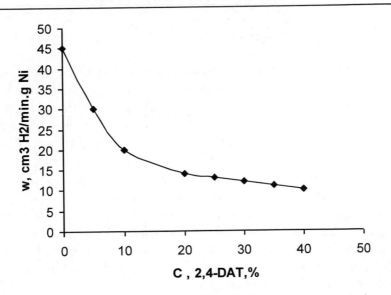

Figure 1. Dependence of the observed rate of 2, 4-DNT hydrogenation upon the content of 2, 4-DAT.
Conditions: temperature 398K, the catalyst quantity is 1 g, hydrogen pressure: 1 atm, the size of catalyst particles: 20 nm, solvent: ethanol-water 80:20 mass%.

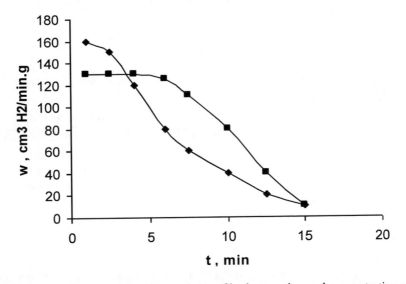

Figure 2. The dependence of the observed rate of hydrogenation and concentration of 2, 4-DNT upon the reaction time.
Conditions: ethanol-water-2, 4-DAT 50:12:38 mass % - temperature:298K, catalyst quantity: 0.625 g, hydrogen pressure: 1 atm, catalyst particle size: 20 nm.

3.1. The Effect of Hydrogen Pressure

The study of the effect of pressure changing upon the rate 2, 4-DNT hydrogenation was done in a region corresponding to the zero order reaction for the majority of hydrogenation reaction proceeding in region of a zero-order with respect to the hydrated compound there was an observed diffusion retarding the reaction on hydrogen. In these studies the conditions of 2, 4-DNT hydrogenation were chosen so that reaction corresponded to the kinetic region. It is seem from the comparison of Till parameter values calculated according to the data of the kinetic experiment.

The observed rate of 2, 4-DNT hydrogenation was 120 cm^3 H_2 / min.g Ni taking account the catalyst rate it will correspond to the hydrogenation rate on the surface of a catalyst equal to 500 cm^3 H_2 / min.g Ni (Figure 2). The value of the rate constant for 2, 4-DNT hydrogenation reaction taking into consideration the stability of hydrogen will be:

$$K^0_s = w^0_s / C^0_{H2} = 500 / (61.5 \times 10^2) = 200 \ c^1 \tag{1}$$

and the value of Till modules in this case will be:

$$F = k^0_s \times R^2 / D^* = 200 \times (2.5 \times 10^{-4}) / 2 \times 10^{-5} = 0.40 \tag{2}$$

Where, D^* is the effective hydrogen diffusion coefficient on the catalyst pores, R is the radius of the particles for Raney – Ni. The results of kinetic study shown in Figure 3 confirm the fact that 2, 4-DNT hydrogenation by Raney – Ni nanoparticles in the conditions chosen proceeds in the kinetic region.

When hydrogen pressure increases to 1.5 atm, the rate of 2, 4 –DNT hydrogenation rate increases linearly (Figure 4), then this linearity is destroyed and when the pressure reaches 11.5 atm the rate of 2, 4-DNT hydrogenation ceases to the dependent upon hydrogen pressure. Similar character of the dependence of nitro compounds hydrogenation rate upon hydrogen pressure has been discussed.

The effect of hydrogen partial pressure on the initial rate of hydrogenation and average catalytic activity was also studied. The initial rate of hydrogenation increased with hydrogen pressure and thereafter it remained unchanged. The average catalytic also followed a similar trend.

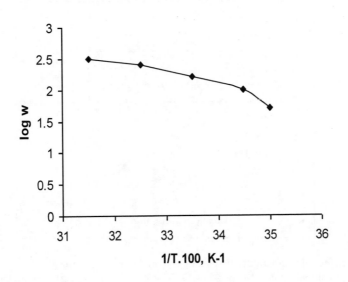

Figure 3. The dependence of 2, 4-DNT hydrogenation rate upon the reaction temperature in the coordinates of Arhenius equation.
Conditions: solvent: ethanol-water-2, 4-DAT – 50:13:38 mass %, catalyst particle size: 20 nm, catalyst amount 1 g., 2, 4-DNT quantity: 1 g.

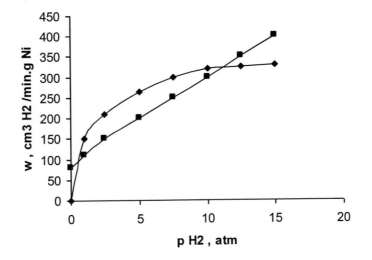

Figure 4. The dependence of the observed rate of 2, 4-DNT hydrogenation reaction upon hydrogen pressure and Langmuir equation in linear coordinates.The experimental conditions correspond to the conditions in Figure.

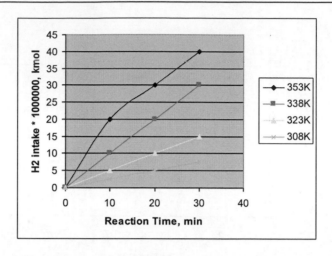

Figure 5. Effect of temperature on hydrogen consumption profiles.
Conditions: solvent: ethanol-water-2, 4-DAT – 50:13:38 mass %, catalyst particle size:
20 nm, catalyst amount 1 g., 2, 4-DNT quantity: 1 g.

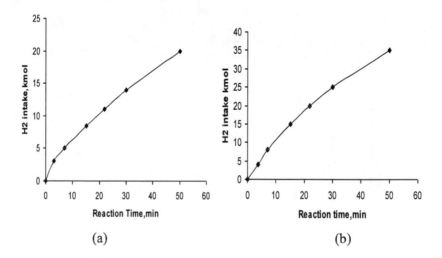

(a)   (b)

Figure 6. Effect of catalyst particle size on the initial hydrogenation rate. (a) Micro
particle size catalyst (b) Nano particle size catalyst.

3.2. The Effect of Temperature

The temperature dependence of rate constant for 2, 4-DNT hydrogenation
has two linear plots with the bending in the region of temperature equal 298K
(Figure 5). The values of the activation energy calculated according to the data

in Figure 3 for the first and second plots are 25 and 54 kJ/mol, respectively. The kinetic field of 2, 4-DNT hydrogenation will confirm to the following condition of carrying out the process: the reaction temperature should be not higher than 353K, the hydrogen pressure should be about 1 atm, the size of the particles of catalyst should be not more than 20 nm and the reaction rate on the catalyst surface is 500 cm^3 H_2 / min.g Ni.

3.3. Effect of Particle Size Catalyst

Figure 6 shows the activity of Raney-Ni nanoparticles in the micro and nano sizes for the hydrogenation of 2, 4-DNT to 2, 4-DAT in ethanol(90%) solvent was investigated at 353K and 1 atm of hydrogen pressure and the results are presented in Table 1.The results show that the initial hydrogenation rate and catalytic activity increasing for Raney-Ni nanoparticles.

Table 1. Effect of transition metal catalyst particle size on the initial rate of hydrogenation and catalytic activity of hydrogenation of 2, 4-DNT

Samples	Catalyst	$R_A \times 10^3$ (kmol/m^3.sec)	N (kmol/kg.hr)
1	(Micro size) Ni-Al	1.658	1.548
2	Ni-Al(Nano size)	3.654	2.954

Reaction conditions: 2, 4-DNT 0.480 mol/cm^3, ethanol(90%) 30ml, catalyst: 0.266 g/cm^3, H_2 pressure:1 atm, temperature: 353K, agitation: 1000rpm.

Table 2. Effect of solvent on initial rate of hydrogenation and catalytic activity during hydrogenation of 2, 4-DNT

Samples	Solvent	Dielectric constant)ε(	$R_A \times 10^3$ (kmol/m^3.sec)	N (kmol/kg.hr)
1	Water	78.5	1.254	2.987
2	Methanol	32.6	1.210	1.874
3	Ethanol	24.3	1.023	1.658
4	n-Prpanol	20.1	0.687	1.547

Reaction conditions: 2, 4-DNT 0.480 mol/cm^3, catalyst: 0.266 g/cm^3, H_2 pressure:1 atm, temperature: 353K, agitation: 1000rpm.

3.4. Effect of Solvent

The effect of solvents on the initial rate of hydrogenation and average catalytic activity was also investigated using Raney-Ni nanoparticles at 353K. The solvents used for this study include water, methanol, ethanol, n-propanol. The results are presented in Table 2.

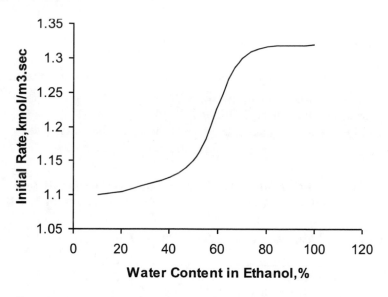

Figure 7. Effect of water content in ethanol on initial rate of hydrogenation and catalytic activity.

The average catalytic activity and initial rate of hydrogenation was highest in water as a solvent and lowest for n-propanol. It was observed that both, hydrogenation rate and catalytic activity increased with increase in polarity of the solvent. This conclusion was further supported by the effect of water content on the catalytic activity and the initial rate of hydrogenation, which increased with water content(Figure 7).

The hydrogenation of 2, 4-DNT is an example of multiphase(gas-liqiud-solid) catalytic reaction system and therefore, it was important to ensure that mass transfer effects were either eliminated or accounted for while determining the instrinsic reaction kinetics. In order to ascertain the importance of mass transfer effects, it was essential to know the solubility of hydrogen in aqueous ethanol. The solubility of hydrogen in ethanol (90%) and Raney-Ni nanoparticles was measured using an absorption technique in a pressure autoclave. In a typical experiment for solubility measurement, a known volume of 90% ethanol (30mL) was charged into the reactor and the contents were heated to a desired temperature. The solubility of hydrogen in ethanol(90%) and Raney-Ni nanoparticles was measured at different partial pressures of hydrogen and temperatures. The solubility of hydrogen was found to increase with hydrogen partial pressure at all the temperatures.

CONCLUSIONS

The hydrogenation of 2, 4-DNT to 2, 4-DAT using Raney-Ni nanoparticles(20nm) was studied in a laboratory scale high-pressure reactor. The experimental data were shown that the initial rates of hydrogenation and average catalytic activity increased with Raney-Ni nanoaprticles. The addition of water resulted in increase in the catalytic activity as well as initial rate of hydrogenation. The kinetics of hydrogenation of 2, 4-DNT to 2, 4-DAT was investigated using Raney-Ni nanoaprticles in a temperature range of 308-353K, hydrogen pressure 1 atm, initial 2, 4-DNT concentration between 0.119-0.958 $kmol/m^3$ and agitator speed between 600-1200rpm.

REFERENCES

[1]   S. Ramyro, V. Gray, W. Scott, Engel, P., Reduction of nitro- nytrozo, azoxy compounds, *J. Org.Chem.* 84 (1989) 4026-4031.

[2]   L. Shricant, R. Karva, Selective catalytic hydrogenation of nitrobenzene to hydrazobenzene *Ind. Chem.Res.* 27 (1989) 21-28.

[3]   L.K. Petrov, K. Kumbilieva, N. Kirkov, Kinetic model of nitrobenzene hydrogenation to aniline over industrial cooer catalyst considering the effects of mass transfer and deactivation *Catalysis* Amsterdam 59 (1990) 31-40.

[4]   H. Smith, W. Bediot, Kinetics of the catalytic hydrogenation of the nitrogroup on Pt *J. Phys. Coll. Chem.* 17 (1951) 1085.

[5]   T.A. Palchevskai, L.B. Bogytskai, B.M. Belocov, Hydrogenation of substitiuted aromatic nitro compounds over Pd/C *J. Chem.* Okrania 58 (1990) 1428-1436.

[6]   G. Horanyi, G. Vertes, Study of p-nitriphenol on platinized platinum electrodes", *J. Electroanal. Chem.* 43 (1973) 441-450.

[7] O.M. Kut, F. Yucelen G. Gut, Selective liquid-phase hydrogenation of 2, 6-dinitrotoluene with platinum Catalysts, Catalysts *J. Chem. Biotechnol.* 39 (1987) 107-114.

[8] C.F. Melius, I.M. Moscovitz, A.B. Motorola, A molecular complex model for the chemisorptions of hydrogen on nickel *Surface Sci,* 5 (1976) 2179-2186.

[9] I. Sprocchka, M. Hanika, V. Runicka, Diffusion of gases in liquids, I., measurements of the diffusion coefficient of hydrogen in alcohol *Coll. Chem. Communs.* 34 (1969) 3145-3152.

[10] W. Scifert, P. Condit, selective catalytic hydrogenation of the nitroolefins, *J.Org. Chem.* 28 (1963) 265-271.

[11] H.D. Borah, D. Prajapati, J. sandhu, A. Ghosh, A. Bismuth chloride – zinc promoted selective reduction of aromatic nitro compounds to azoxy compound *Tetrahedron Lett.* 35 (1994) 3164-3169.

[12] J. Freel, W.J.. Pieters, R.B. Anderson, The structure of raney nickel *J. Catal.,* 16 (1969) 281-288.

[13] S.D. Mikhailenko, The effect of redox treatment on the structural, adsorptive and catalytic properties of Raney nickel *J.Catal.* 141 (1993) 688-694.

[14] W.G. Amett, Solvent effects inorganic chemistry *J.Am. Chem.* 87 (1965) 1541-1548.

[15] V.P. Gostikin, K.N. Belongov, J.T. Nikckolayev, Effect of dispersion of Raney nickel on the reduction rate of nitrophenol in liquid phase *Chem.Ind.* 6 (1987) 418-423.

[16] L.K. Filippenko, V.P. Gostikin, L.K. Popov, Effect of the temperature and hydrogen pressure on the rate and the selectivity of the reduction process of 2-nitro-oxy-5-methylbenzene by hydrogen in a liquid phase on the promoted nickel *Catalyst. Chem.Chem.Technol* Russia 29 (1986) 52-58.

[17] S. Nishimura, Handbook of Heterogeneous Catalytic Hydrogenation for Organic Synthesis, Wiley-Interscience, (2001).

[18] G. Kabalka, R.S. Varma, Comprehensive Organic Synthesis, Eds. Trost B.M. F Eming, I. *Pregamon Press*, Oxford, 8 (1991) 363.

[19] P. Tundo, Continuous Flow Methods in Organic Synthesis New york (1991).

[20] S. Saaby, K.R. Knudsen, M. Ladlow, S.V. Ley, *Chem. Commun.* (2005) 2909-2911.

In: Modern Nanochemistry   ISBN: 978-1-61209-992-7
Eds: A. K. Haghi and G. E. Zaikov   © 2011 Nova Science Publishers, Inc.

Chapter 14

CONJUGATED DIENES MEDIATED BY ORGANOSELENIUM NANO-SCALE REAGENTS

M. R. Saboktakin[1] and A. K. Haghi[2]

[1]Department of Nanotechnology, Baku State University, Azerbaijan
[2]University of Guilan, Iran

ABSTRACT

The use of organoselenium reagents in organic synthesis is common practice for functional group manipulation since they generally require mild reaction conditions and afford good yield. In this research, we have developed efficient catalytic processes using p-nitrophenyl selenenyl chloride nanoparticles to optically active products using the chemistry outlined above. The selenenylation can be followed by oxidative β-elimination to regenerate double bonds, which are functionalized in the allylic position. As the first part of a continued research, we have synthesized the p-nitrophenyl selenenyl chloride nanoparticles with high yield.

1. INTRODUCTION

Good yields uniform compounds are two important demands for reactions to be useful in synthetic organic chemistry [1, 2]. Thus, many remarkable efforts have been provided so far to many important transformations amenable to catalysis [3]. Several compounds remain that are not readily obtained, for example, the products of functionalizations of not activated C=C bonds [4, 5]. Various research groups including ourselves have already proven the efficient use of stoichiometric amounts of selenium organic nanomaterials in selective additions [6]. Products of this type were obtained with high levels of selectivity and can be employed as ideal precursors to further transformations leading to compounds by oxidative elimination, to radicals with a rich subsequent chemistry and by further nucleophilic displacement to compounds [7, 8]. The results achieved so far are not discussed as they have been published in recent reviews [9, 10]. The several compounds containing allylic groups are easily accessible by this method. After selenylation of dienes allylic selenides are obtained, the double bond with the higher electron density being more reactive in the selenenylation [11]. These compounds serve as ideal substrates in subsequent rearrangements to yield allylic alcohols as the final products. The rearrangements are involved in the catalytic selenium dioxide oxidations mentioned above as early investigations have shown [12]. In the rearrangement an easer of seleninic acid is formed which has to be cleaved to yield product. In this research, we are targeting the developing of efficient catalytic addition-elimination sequences and selenenylation rearrangements according to the original reaction [13, 14]. To achieve these important goals several issues have to addressed:

1. The generation of the selenium electrophiles by an oxidant, the addition and the subsequent oxidative elimination / rearrangement conditions must be compatible to each other.
2. The oxidative elimination as well as the rearrangement cleaves off the selenium reagent, which must either be converted to or be able to react again as an eletrophile directly.
3. The double-bond in the product must be less susceptible for a next addition reaction than the double-bond of the starting material [15].

2. MATERIALS AND METHODS

The red elemental selenium used was made by John Coffman procedure. In this method, first, 50ml of 2M ascorbic solution (laboratory grade reagent) was mixed with 50 ml of 0.2M NaOH solution to make 0.1 sodium ascorbate solution. Second, 25 ml of 0.1M sodium biselenite solution was added to this clear salt solution without stirring. The pH value was adjusted to 2.5 by adding concentrated hydrochloric acid and the solution appeared red at that time. Third, this solution was centrifuged at 15000 rpm for 20 min. after separation, the red precipitate was washed by DI water two time, then was lyophilized by liquid nitrogen and collected. The prepared red selenium was heated in DI water for 30 min until grey / black allotrope selenium was appeared. Then, the grey / black selenium was dried in room temperature. 2.3 g metallic sodium is slowly added under nitrogen to a suspension of 7.9 g of grey/black selenium in 250 ml hydrazine hydrate and sodium hydroxide at -40 °C. After addition, the solution is stirred for 1 hour, then the product is dried by a sublimation of the water component in an iced solution (yield 4.3g, 64.4%). To 375 ml (4.3moles) of hydrochloric acid (sp. gr. 1.18) and 200 g of ice is added 139.6(1.5moles) of p-nitroaniline. The result solution is diazotized with a solution of 103.5 g (1.5moles) of c.p. sodium nitrite, ice being added to the reaction mixture, as necessary, in order to keep the temperature below 5 °C. The final volume of diazotized solution is about 1L. This solution is added in a slow stream from a dropping funnel to the sodium diselenide solution, which is vigorously stirred with a mechnical stirrer. The diazotized solution is heated to boiling. It is then poured back on the oil, the mixture is well stirred, 200ml of chloroform is added and the selenium is collected on a filter and washed with a little more chloroform. After the chloroform layer is separated, the aqueous layer is again extracted with 200 ml of chloroform. The combined extracts are then distilled, the p-nitro-diphenyl diselenide being collected. A 1 L three – necked, round-bottomed flask equipped with a magnetic stirring bar, a thermometer, a gas – inlet tube, and a reflux condenser is charged with 50 g (0.16 mole) of p-nitro-diphenyl diselenide and 350 ml of hexane. The mixture is warmed to 40-50°C, dissolving the solid. The resulting solution is stirred while 2.1g thionyl chloride and a little dimethyl formamide was added to it. The solution is heated to reflux for 1 hour, filtered by gravity, and allowed to cool slowly at room temperature then at 6 °C. The mother liquor is decanted, the large, deep- orange crystals are washed with 25ml of pentane. After approximately 180 min, the solution was sprayed into a liquid nitrogen bath cooled down to 77° K, resulting in frozen droplets. These frozen droplets were then put into the chamber of the freeze-dryer. In the freeze-drying process, the products are dried by a sublimation of the water component in an iced solution., giving 51.6 g(96.5%) of p-nitro-phenyl selenenyl chloride. The reaction of 1.1 equiv. of p-nitrophenyl selenenyl chloride in the presence 1.1 equiv. of the hindered base 2, 6-di-tert-butyl-4-methylpyridine(DTBMP) in dichloromethane solution was conducted in the -78 °C for 3 hour. Under these conditions, complete consumption of the starting material was observed and a 9:1 mixture of product was produced (yield 98.2%).

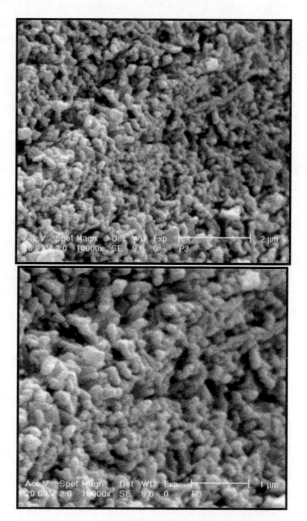

Figure 1. SEM of p-nitro-diphenyl diselenide nanoparticles (20-30nm).

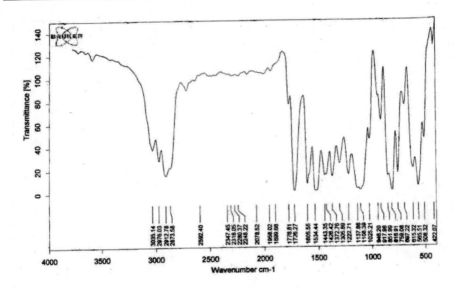

Figure 2. FT-IR spectrum of p-nitrophenyl selenenyl chloride (1).

3. RESULTS

Figure 1 illustrates the micrograph of the obtained p-nitrophenyl selenenyl chloride nanoparticles (20-30nm) after freeze-drying method. In this research, we have discussed the effect of selenium organic compound particle size on the reaction yield. Therefore, The results show that selenium organic compound particles size increase reaction yield and conversion of products.

Figure 2 shows the FT-IR spectrum of p-nitrophenyl selenenyl chloride (1), where the % of transmittance is plotted as a function of wave number (cm^{-1}). The characteristic FT-IR peak at 3035 and 1605, 1443 cm^{-1} are due to the presence of =CH and C=C aromatic ring, respectively. Also, The peaks at 1590, 1312cm^{-1} are due to the NO$_2$ bond stretching vibration.

Figure 3 shows the 400MHz ^{1}H-NMR spectrum of p-nitro diphenyl diselenide(1) into chloroform solvent(CDCl$_3$). Integration of the CH$_2$ doublet at 3.33-3.48 ppm. Integration of phenyl hydrogens at 7.46-7.70 ppm. The sample was held at 20°C for 12h before recording the NMR spectra.

Figure 4 shows the FT-IR spectrum of 3-Methyl-2-buten-1-ol (2), where the % of transmittance is plotted as a function of wave number (cm^{-1}). The characteristic FT-IR peak at 3620, 1605, 1108 cm^{-1} are due to the presence of OH, C=C aromatic ring, C-O, respectively.

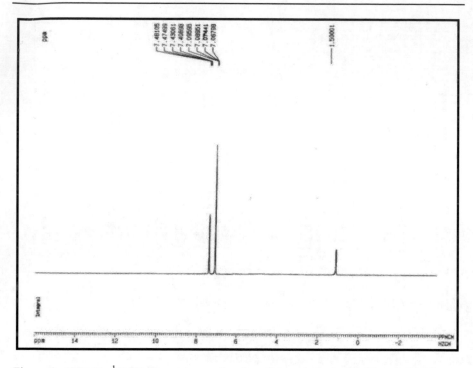

Figure 3. 400MHz ^{1}H-NMR spectrum of p-nitro diphenyl diselenide(1) into CDCl$_3$.

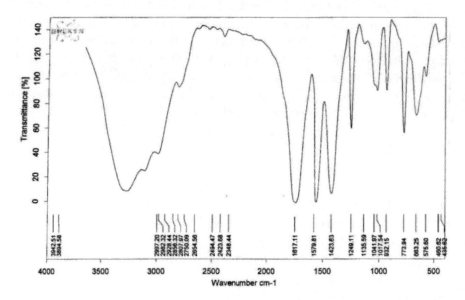

Figure 4. FT-IR spectrum of 3-Methyl-2-buten-1-ol (2).

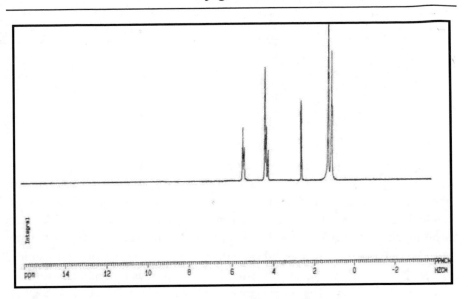

Figure 5. 400MHz ^{1}H-NMR spectrum of 3-Methyl-2-buten-1-ol (2) into CDCl$_3$.

Figure 5 shows the 400MHz ^{1}H-NMR spectrum of 3-Methyl-2-buten-1-ol (2) into chloroform solvent (CDCl$_3$). Integration of the CH$_3$ triplet is at 5.42ppm, CH$_2$ doublet is at 4.25 ppm, CH singlet is at 2.21 ppm and CH$_2$OH doublet is at 1.82 ppm. The sample was held at 20°C for 12h before recording the NMR spectra.

4. DISCUSSION

The synthesis of selenium organic compounds were approached with protected p-nitroaniline. Our initial route to these compounds focused on modification to literature preparation of selenium organic and the corresponding selenium chloride derivatives. p-nitroaniline was converted into diazonium salt, which was treated with Na$_2$Se$_2$ to afford the protected p-nitrodiphenyl diselenide.According to general method, Compound 2 was treated with thionyl chloride in Dimethyl formamide and reflux for 1 hour to convert the diselenide function into the corresponding selenenyl chloride nanoparticles(1). All analytical data obtained for this compound are in accordance with the data reported in the literature.

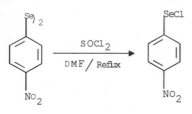

(1)

From these results, it is clear that p-nitrophenyl selenenyl chloride nanoparticles may be an ideal starting material for the preparation of substituted selenium reagent derivatives. By employing the above mentioned methodology, other diselenenyl derivatives was transformed into the corresponding selenenyl chloride. The subsequent reaction with benzene selenolate (PhSe⁻Na⁺), generated in situ by reducing diphenyl diselenide (PhSeSePh) proved to be an effective way to produce phenyl substituted selenenyl derivatives. The improvement of yield in this reaction by the addition of NF₃ led to an assumption that a selenium moiety containing an internal material would provide an interesting strategy to synthesize selenenyl derivatives containing heteroatoms in close proximity to selenium.On of the important objectives of this work was to study efficient nanomaterials to reaction yield using the chemistry outlined above. The selenenylation can be followed by oxidative β-elimination to regenerated double bonds, which are functionalized in the allylic position.We have already shown that this sequence can be performed with very efficiency using diselenides as outlined below. In this step, we have used from 1, 3-pentadiene as initial material. Compound 3 cotaining allylic groups are easily accessible by this method. After selenenylation of 1, 3-pentadiene, allylic selenides of type 2 are obtained, the double bond with the higher electron density being more reactive in the selenenylation. This compound serve as ideal substrates in subsequent rearrangements to yield allylic alcohol 3 as the final product. The re-arrangements are involved in the catalytic selenium dioxide oxidations mentioned above as early investigations have shown. In the rearrangement an ester of seleninic acid is formed which has to be cleaved to yield product 3.

In this research, we are targeting the development of yield of addition – elimination sequences and selenenylation – rearrangements according to the two reaction have shown above using selenium organic nanomaterials. To achieve these important goals several issues have to be addressed.

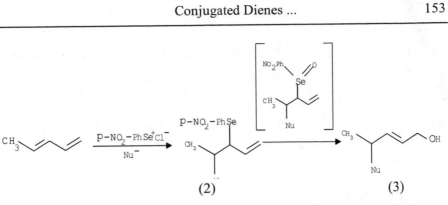

(2)   (3)

CONCLUSIONS

On of the important objectives of this work was to study the effects of organic selenium nanomaterials (20-30nm) on the selenenylation of conjugated dienes. The selenenylation can be followed by oxidative β-elimination to regenerated double bonds, which are functionalized in the allylic position. We have already shown that this sequence can be performed with very efficiency using diselenides using p-nitrophenyl selenenyl chloride nanomaterials as outlined below. These results show that the p-nitrophenyl selenenyl chloride nanomaterials, effectly, can be increase of the selenenylation of conjugated dienes.

REFERENCES

[1]   Denmark, S.E.; Fujimori, S.; "In modern aldol reactions", *Mahrwald, R., Ed.; Wiley-VCH: Weinheim, 2004, 2, 7.*

[2]   Yamamoto, H.; Oshima, K.; "Main group metals in organic synthesis"; *Ed.; Wiley-VCH: Weinheim, 2005, 1, 7.*

[3]   Akiba, K.; "Chemistry of hypervalent compounds"; *Ed.; Wiley-VCH: Weinheim, 1999, 2, 7.*

[4]   Finet, J.P.; "Ligand coupling reactions with heteroatomic compounds"; *Pergamon:Oxford, 1998.*

[5]   Jacobsen, E.N.; Pfaltz, A., Yamamoto, H.; "comprehensive asymmetric catalysis"; *Ed.; Wiley-VCH: Weinheim, 2000, 1, 1.*

[6]   Brunner, H.; Zettlemeier, W.; *Ed.; Wiley-VCH: Weinheim, 1993, 1, 2.*

[7]   Beller, M.; Bolm, C.; "Transition metals for organic synthesis"; *Ed.; Wiley-VCH: Weinheim*1998, 1.

[8]   Paulmeir, C.; "Selenium reagents and intermediates in organic synthesis"; *Pergamon Press, Oxford,* 1986.

[9]   Tiecco, M.; "Electrophilic selenium, selenocyclization, in Topic in current chemistry:organoselenium chemistry"*springer, Heidelberg*, 2000, 7-54.

[10]   Nicolaou, K.C.; Petasis, N.A.; "Selenium in natural product synthesis"; *CIS*: Philadelphia, 1984.

[11]   Back, T.G.; "In the chemistry of organic selenium and tellurium compounds"; *Patai, S., Ed.; Wiley:New York, 1977, supplement A*, part 2, 854-866.

[12]   Schmid, G.H.; Garratt, D.G.; "In the chemistry of double bonds functional groups "; *Patai, S., Ed.; Wiley:New York, 1977, supplement A*, part 2, 854-866.

[13]   Terret, N.K.; "In combinational chemistry"; Oxford university Press, Oxford, 1998.

[14]   Fields, G.B.; "Solid – phase peptide synthesis.In methods in enzymology"; *Academic Press, san diego*, 1997.

[15]   Urgess, K.; Van der Donk, W.; "In advanced asymmetric synthesis "; *Stephenson, Ed.; Chapman & hall*:London, 1996.

INDEX

J

K

I

L